Introduction to Symbolic Plan and Goal Recognition

Synthesis Lectures on Artificial Intelligence and Machine Learning

Editors
Ronald Brachman, *Jacobs Technion–Cornell Institute at Cornell Tech*
Francesca Rossi, *IBM Research AI*
Peter Stone, *University of Texas at Austin*

Series Page

Introduction to Symbolic Plan and Goal Recognition
Reuth Mirsky, Sarah Keren, and Christopher Geib

    ISBN: 978-3-031-00461-2   paperback
    ISBN: 978-3-031-01589-2   ebook
    ISBN: 978-3-031-00034-8   hardcover

    DOI 10.1007/978-3-031-01589-2

A Publication in the Springer series
SYNTHESIS LECTURES ON ARTIFICIAL INTELLIGENCE AND MACHINE LEARNING

Lecture #47
Series Editors: Ronald Brachman, *Jacobs Technion–Cornell Institute at Cornell Tech*
               Francesca Rossi, *IBM Research AI*
               Peter Stone, *University of Texas at Austin*
Series ISSN
Synthesis Lectures on Artificial Intelligence and Machine Learning
Print 1939-4608   Electronic 1939-4616

Introduction to Symbolic Plan and Goal Recognition

Reuth Mirsky
University of Texas at Austin

Sarah Keren
Harvard University

Christopher Geib
SIFT

SYNTHESIS LECTURES ON ARTIFICIAL INTELLIGENCE AND MACHINE LEARNING #47

ABSTRACT

Plan recognition, activity recognition, and goal recognition all involve making inferences about other actors based on observations of their interactions with the environment and other agents. This synergistic area of research combines, unites, and makes use of techniques and research from a wide range of areas including user modeling, machine vision, automated planning, intelligent user interfaces, human-computer interaction, autonomous and multi-agent systems, natural language understanding, and machine learning. It plays a crucial role in a wide variety of applications including assistive technology, software assistants, computer and network security, human-robot collaboration, natural language processing, video games, and many more.

This wide range of applications and disciplines has produced a wealth of ideas, models, tools, and results in the recognition literature. However, it has also contributed to fragmentation in the field, with researchers publishing relevant results in a wide spectrum of journals and conferences.

This book seeks to address this fragmentation by providing a high-level introduction and historical overview of the plan and goal recognition literature. It provides a description of the core elements that comprise these recognition problems and practical advice for modeling them. In particular, we define and distinguish the different recognition tasks. We formalize the major approaches to modeling these problems using a single motivating example. Finally, we describe a number of state-of-the-art systems and their extensions, future challenges, and some potential applications.

KEYWORDS

plan recognition, goal recognition, activity recognition, behavior recognition, intent recognition, temporal pattern recognition, reasoning under uncertainty, human–AI collaboration, multi-agent systems, symbolic reasoning

To Ilya, for giving me something greater than research, and then letting me have both.

– Reuth Mirsky

To my husband Uri and to my Ph.D. advisors, Avigdor Gal and Erez Karpas. I couldn't have done this without your support.

– Sarah Keren

To Sabra. Shockingly, sometimes words actually do fail me.... R&C.

– Christopher Geib

Contents

Preface

This book is based on a tutorial presented by the authors at the 2019 AAAI conference in Hawaii. The wish to create the tutorial came up after the authors had been co-chairing the Plan Activity and Intent (PAIR) workshop at AAAI since 2017. The PAIR workshop, initially named Modeling Others from Observations (MOO), has been taking place at different venues since 2004. Every year, it attracts and brings together researchers from diverse backgrounds and disciplines.

Our experience as co-chairs helped us appreciate the increasing interest in plan and goal recognition. However, it also highlighted the lack of a shared vocabulary and terminology to connect the different lines of work. Our intention for the tutorial and this book is to bridge the gap that exists between different threads of research in the field. We do this by providing an overview of past and state-of-the-art model-based plan and goal recognition literature, by specifying a formalization of the elements of the problem, and by describing a set of practical tools for evaluating and investigating a new recognition problem.

The book is organized into five chapters. Chapter 1 is an introduction to plan and goal recognition and an overview of key past works in the area. Chapter 2 provides a unified recipe for defining a recognition problem, and provides guidelines for choosing an approach for a given recognition task. In Chapter 3, we formalize goal and plan recognition. Chapter 4 describes a variety of state-of-the-art approaches to recognition and suggests ways to extend existing symbolic plan and goal recognition tools. Finally, Chapter 5 highlights possible directions for future work. In all of these chapters we provide references an interested reader can use to continue the exploration of this research space.

We hope this book will enable and encourage researchers to read more widely past work and to build on its lessons to advance plan and goal recognition research. To paraphrase Sir Isaac Newton, our research field truly has been built on the shoulders of giants.

Reuth Mirsky, Sarah Keren, and Christopher Geib
January 2021

Acknowledgments

While the cover of this book lists three authors, no such work is ever completed without the assistance, encouragement, and support of a strictly non-enumerable number of supporters, colleagues, editors, mentors, friends, and family (genetic and chosen). We thank you all. Further, we would like to acknowledge all of the exceptional researchers whose work we reference here. We have worked to the best of our ability at synthesizing a diverse field of research, and if in that effort our characterization of your work has been imperfect, we apologize. We would also like to acknowledge all of the authors of papers at the PAIR and MOO series of workshops that has been such a rich and supportive research community over the years and contributed to and shaped so much of our thinking about these problems. Gal Kaminka the "founding father" of the first MOO workshop deserves special acknowledgment in this regard. We would also like to thank all of the people that encouraged us (before and after) to run the Plan Activity and Intent Recognition Tutorial at AAAI-19 that was the impetus behind this volume.

We would all like to thank Michael Morgan and the whole team at Morgan & Claypool for their patience and support, Peter Stone for suggesting we turn our tutorial into a book, and our reviewers Gita Sukthankar, Shlomo Zilberstein, and David Pynadath for their comments on an earlier draft of this book. It has benefited immeasurably from their efforts and any errors in it were definitely inserted after their commentary. We would like to acknowledge and thank our partners and families for their patience and support during this process. Writing a book is never an easy or fast process and it is not only the authors that pay the costs. Thank you.

More specifically, Reuth would like to thank Kobi Gal for introducing her to the world of PAIR; Chris Geib and Gal Kaminka for being the Gandalf and the Elrond of her first PAIR adventures; Sarah Keren, Mor Vered, and Maayan Shvo for exciting discussions about PAIR and beyond; and Peter Stone for his inspiring mentorship and support.

Sarah would like to thank Avigdor Gal and Erez Karpas, her Ph.D. advisors who helped formulate Goal Recognition Design and introduce it to the world. She also thanks Barbara J. Grosz, David C. Parkes, and Jeffrey S. Rosenschein for supporting her in the process of writing this book and for helping her find new and better ways to make her messages clear.

Finally, Chris would like to thank his co-authors Reuth and Sarah. It is in our discussions, conversations, arguments, and efforts to write down what we know that we clarify our own knowledge. The benefits of having you as such insightful, challenging, and thought-provoking, colleagues has been immeasurable. Thank you.

Reuth Mirsky, Sarah Keren, and Christopher Geib
January 2021

CHAPTER 1

Introduction

One of the questions asked by Artificial Intelligence (AI) researchers for the last 60 years has been, "What actually constitutes or defines 'intelligence'." The proposal for the 1956 Dartmouth workshop that coined the term AI explicitly mentions natural language processing, machine learning, and general problem solving as requirements of intelligence. Since that time a number of other AI problems have been pointed out as core to our understanding of human-level intelligence including: planning and scheduling, game playing, vision, and other perceptual tasks. However, we would claim there is at least one more that has received less attention, namely plan, activity, and intent recognition.

Twelve years before the Dartmouth workshop, Heider and Simmel [1944] showed that human subjects, when shown a movie of nothing more than black triangles and circles moving on a white field, not only characterized the triangles and circles in anthropomorphic terms, but explicitly ascribed to them intentions and plans, and described their motion as specific goal-directed actions. That is, the people watching the films first ascribed agency to the moving shapes and then recognized their actions as being part of temporally extended plans to achieve specific goals. To this day, people viewing the film for the first time are surprised by the strength of their own immediate impressions.

More recently, Warneken and Tomasello [2007, 2009] and Knudsen and Liszkowski [2012] have argued that pre-linguistic children will not only recognize the plans of adults they do not know, but on recognizing a plan failure will act to help the adult achieve their plan and goal. Keep in mind that to do this, the children must recognize the adult's goal, the structured plan to accomplish the goal, where the adult is within the plan's execution, and the manner in which the plan is failing. The child must then recognize that they have the ability to take action to either fix the plan or at least gesture to warn the adult of its impending failure. All of which, these children are doing before acquiring full linguistic proficiency. All of which, will be critical to achieving any of their own goals over the course of their lives. All of which, we would characterize as *intelligent*.

This said, regardless of one's views on recognizing plans and goals of other agents as a marker of intelligence, there are many tasks we would like to have computers perform that require recognizing the plans and goals of others. For example, plan recognition is critical to natural language processing. Consider the following interaction.

Scenario: State of the art

Setting: music playing in the background
Bob: "Siri, What song is playing?"
Siri: "The song currently playing is 'Long Time Gone' by the Chicks."

Siri and other current generation "personal assistants" are able to answer specific factual questions of the kind one can rewrite into a simple logical form (usually conjunctive) and put directly into any current internet search engine, and nothing more.

Now consider what a human personal assistant, Jeeves, might say in this setting.

Scenario: Desired

Setting: music playing in the background
Bob: "Jeeves, What song is playing?"
Jeeves: "This song is 'Long Time Gone' by the Chicks. If you are interested, the band has just released their first album in more than ten years, and will be playing locally in two months. Would you like me to try to get you tickets for the show?"
Bob: "Are Reuth, Sarah, and Chris free?"
Jeeves: "I'll find out if they would like to join you, sir."

Jeeves infers that if Bob liked the song well enough to ask about it, he might like to see the band live. In furtherance of that recognized goal, Jeeves suggests a plan that he could help with by asking if Bob would like tickets to the Chicks' next show. Far more difficult for current systems, when asked a yes or no question about tickets, rather than answering, Bob asks another question about the availability of other people. For this to make sense as a response, Jeeves must recognize that Bob is considering a more complex plan than the one suggested and includes asking the authors of this book to join him in attending the concert. Having realized this, Jeeves mentally modifies his proposed plan, and acknowledges the changed plan by indicating that he will take action to find out the information requested and negotiate a potential date where all can attend.

Such reasoning is far beyond state of the art in personal assistant systems precisely because they don't include plan recognition both at the discourse/language level (i.e., how to understand answering a question with a question) and at the level of executing actions to attend the concert (i.e., inferring from a question of interest to a desire to see a band as well as other actions that would help Bob achieve his goals). Human users would very much like to have assistant systems that are able to help them in this manner, but current systems are simply not up to the task. For example, the failure of the Microsoft help assistant "Clippy" to recognize when its "help" was inappropriate made it much more annoying than helpful, and led to its removal from the Microsoft tool suite.

There are many other real-world application areas for plan recognition, including intelligent tutoring and educational systems, recognizing the plans and intent of hostile "hackers" in cyber security settings, developing believable characters in video games and entertainment, automated control systems for large-scale manufacturing, and almost any sufficiently complex user interface to a computerized system. We have written this book because of our interest both in the foundational need for an understanding of plan recognition as part of intelligent systems as well as the critical need for it in real deployed systems. We believe plan, activity, goal, and intent recognition are the critical next research problems that need to be solved for our systems to make the next step toward achieving the long sought-after goal of systems that adapt to their users rather than forcing users to adapt to their systems.

1.1   SOME NECESSARY REMARKS ON SCOPE, PRE-HISTORY, AND TAXONOMY

This book attempts to bring together and provide an overview of the current state of work in AI on plan, activity, and intent recognition. However, depending on one's definitions and the lens though which work is viewed, there is a vast amount of research that might be considered within this scope. This means, our book will, of necessity, be incomplete. Our objective, therefore, will not be to provide a complete treatment of all prior work in the area but rather make high-level distinctions, lay out general research directions, and make clear where we have cut corners or limited our discussion. In these cases we will attempt to provide some citations for a place to start for those interested in these areas not covered. This will also require identifying different subproblems with the field. For example, while there is significant relevant work on Markov Decision Processes and Hidden Markov Models [Howard, 1960] from the 1930s and 1940s that laid the foundations for what we will call *activity recognition*, we will not cover this work in any detail. Instead, we will focus our work within the stream of symbolic AI. To do this will require identifying different subproblems with the field, and to make this as clear as possible it will be helpful to go back and discuss some early history of AI starting with the previously mentioned Dartmouth conference. (See timeline below.)

Early Milestones in AI and Plan Recognition

- 1956 - Dartmouth conference occurs coins the term AI.

- 1959 - General problem-solver (GPS). Newell et al. [1959]

- 1971 - STRIPS planner. Fikes and Nilsson [1971]

- 1978 - Schmidt, Sridharan, and Goodson, "The Plan Recognition Problem: An intersection of psychology and artificial intelligence." Schmidt et al. [1978]

It is worth recognizing that in the mid-1950s many of the distinct subfields of AI that are well known today including planning and scheduling, defeasible reasoning, probabilistic reasoning, knowledge representation, logical inference of almost all kinds, machine learning, and others were viewed as specific instances of "problem solving" and as such were not distinguished as separate fields of study. The prevailing view of the time being that they could all be addressed by correctly representing them and using directed heuristic search. As such, Newell, Shaw, and Simon's General Problem-Solver was intended to be a general-purpose method to address almost any of them including plan recognition [Newell et al., 1959]. Note that continuous sensor integration was generally not seen as part of AI at this time, but would become very relevant to what we call *activity recognition.*

However, with advances in formal theory of computation it was realized that specific subproblems might have different complexities and therefore be amenable to different algorithms or solution methods. As such, the study of specific subproblems might provide insights into the larger problem. This gave rise to both the well-known foundational STRIPS planner by Fikes and Nilson, as well as the first work to discuss plan recognition as such [Schmidt et al., 1978]. In this work, Schmidt, Sridharan, and Goodson put forward the following definition of the plan recognition problem.

> "The problem of plan recognition is to take as input a sequence of actions performed by an actor and to infer the goal pursued by the actor and also to organize the action sequence in terms of a plan structure. This plan structure explicitly describes the goal-subgoal relations among its component actions."

While this definition goes a long way to provide a succinct grounding of the problem, in the 30 years that followed there were two significant realizations. First, like many other AI problems, formalizing it in (a potentially specialized) logic and using search to solve it would not provide algorithms that would run fast enough for real-world use. Second, it was found that a number of different problems were actually at least partially covered by this definition. This meant that a number of qualitatively different, and difficult research problems were explored under the same name, and for a time this led to significant confusion about what problem was actually being addressed when one talked about plan recognition. For the purposes of this book we will separate and distinctly identify three sub-problems that have all been referred to as plan recognition at least at some point: activity recognition, goal recognition, and plan recognition. For these three problems we have developed three related but different icons that we will use in the various sections when we talk about systems to remind the reader what problem the discussed system is attempting to solve. However, as the problems are related the icons attempt to capture their overlapping scope.

- ⊖ *Activity Recognition*: The least abstract level of inference focusing on labeling a sequence of possibly noisy sensor inputs with a single, usually unstructured, label (e.g., "walking," "running," "swimming," "sitting," etc.) denoting a low-level activity. Sometimes also called *behavior recognition*.

- ⊖ *Goal Recognition*: The most abstract level of inference focusing on identifying the highest level goal or overarching objective of the agent's action. More often than not, like activity recognition, this is captured in an unstructured label (e.g., "Making breakfast," "Writing a book," "Refueling a plane," etc.). Note that this means that depending on the level of abstraction and the domain the same task can be described as either goal or activity recognition.

- ⊖ *Plan Recognition*: The recognition of a specific structured plan that is being followed by an agent, including the goal being pursued and where in the execution of the plan the agent is. This is closest to Schmidt et al.'s original definition.

In the following subsections we will more clearly define these problems and provide more discussion of their differences. This will include providing example inputs, outputs, a definition of the central problem addressed by each, and some examples of their application.

1.1.1   ACTIVITY RECOGNITION

⊖ **Definition**: *Given inputs from possibly more than one noisy sensor, produce a (usually unstructured) label from a fixed set that identifies the activity being executed by the observed agent's for given time windows.* This is essentially a labeling task answering the question, for each usually short time interval, what activity is being performed by the observed agent.

Input: A sequence of noisy, continuous sensor inputs over time.

Output: A unique label from a fixed set (or a distribution over the labels) for each temporal sub-sequence.

Central problem: The central problem in this task is dealing with lost and potentially corrupted observations within in a continuous input observation stream. As a result, the algorithm's approaches used to address this are generally in the class of Markov Decision Processes (MDPs) and Hidden Markov Models (HMMs), and other probabilistic Markov models.

Alternative characterization: Classification/labeling of noisy temporal observations.

Example: Video segmentation of football plays (e.g., passing, clearing, throw-in, etc.).

1.1.2   GOAL RECOGNITION

Definition: *Given a sequence of (usually noise free) symbolic observations of actions to produce a (usually unstructured) label from a fixed set that identifies the high-level goal of the observed agent for given time windows.* Notice that this problem is very similar to that of activity recognition. However, more often in the goal recognition case the observations do not have noise or a probabilistic component but the hypothesized goal might. In some cases, goal recognition systems have had a larger set of potential goals than activity recognition systems, however their similarity means that methods for doing activity recognition can sometimes be used for doing goal recognition (and vice versa) simply by modeling more (or less) abstract actions in the domain.

Input: An ordered sequence of discrete symbolic input tokens.

Output: A unique label (perhaps with a probability, perhaps not) for each temporal subsequence.

Central problem: Dealing with evidence for multiple conflicting hypothesis. This means such systems often require a means of weighing evidence from different sources like Bayes nets, Fuzzy Logic, or other probabilistic models of reasoning.

Alternative characterization: Classification/labeling temporal observations where each observation can contribute to many possible labels.

Example: Identifying a computer users high level goals from observing their actions within the user interface (e.g., "searching web," "writing a document," "confused," etc.).

1.1.3   PLAN RECOGNITION

Definition: *Given an ordered set of observed actions execute by an agent and a set of plans to be recognized (including the goal of the plan), infer which plans the agent is executing, where they are in the execution of those plans, and the goal the plan is being executed in furtherance of.* Note that this problem encompasses goal recognition as a sub-problem as it requires identifying the goals of the plans. It also requires producing a structured representation of the plan and where the agent is in its execution. It is frequently argued that activity recognition can be used as input to plan recognition systems.

Input: An ordered sequence of discrete symbolic input tokens.

Output: Complex structure capturing plan being executed. Potentially including abstract tasks that have been done and which are yet to do and traditionally the goal of the plan.

Central problem: Combining sequences of lower-level observations into larger structured patterns (probabilistic or not).

Alternative characterization: Complex temporal pattern prefix matching, complex structured sequence prefix matching.

Example: Identify the plan and goal of cyber intruders and their progress though a network (e.g., Bragging, DoS, espionage, what machines do they attack?).

Notice that the icons we have developed for these three recognition problems attempt to capture the relationships between them. Activity recognition being the lowest level of inference from raw sensor signals to labels, the icon has only the bottom third of the circle. Goal recognition being just the identification of the highest levels of the agent's mental models, the icon only covers the top third of the circle. Since plan recognition does both goal recognition and identifies the plans of the agent, and can be seen as taking activity recognition as an input (and so is discrete from it), the icon covers the top two-thirds of the circle but leaves the bottom third uncolored.

1.1.4   THEN WHAT IS "INTENT RECOGNITION"?

Finally, *intent recognition* is a term that has been used on and off in the past, and is now used with much less specificity than these other terms. As a result, we will not use this term in the remainder of the book. When a paper uses this term we would strongly urge the reader to try to identify which of the previous three terms describes the problem they are actually addressing. Early on, this term was synonymous with goal recognition, but unfortunately there is not a consensus on this usage. Some of the things the term *intent* has been used to mean by various groups of people working in and around this general research area are:

Non-AI researchers: Ineffable magic that differentiates the motivations of humans from that of synthetic agents' actions.

Some philosophers: A separate mental pro-attitude toward an action or plan denoting a commitment to its execution [Bratman, 1987].

Other philosophers: A mental state in which an agent both believes a sequence of actions will cause a state they desire and that they will execute those actions to that end [Cohen and Levesque, 1990].

Military: What the most Sr. officer wanted to have happen regardless of the orders given or the actual outcome of the plans.

Given this ambiguity in usage, we will not use this term for the remainder of the book, and we suggest the reader remove it from their use in this context as well.

1.2   THE SCOPE OF THIS BOOK

With these definitions in hand we are now in a position to more closely define the scope of this book. As we have already alluded, this book will not even cover all of the approaches to problems that we have outlined above. With the recent successes of big data (BD), neural networks (NN), and machine learning (ML), such methods have been successfully applied to the problems of goal and activity recognition. However, while such "non-symbolic" approaches have enabled the building of working demonstrations within fixed domains, at this point they are limited both in terms of scale and coverage. The often-touted strengths of such approaches, namely their abilities as arbitrary function approximators and distributed "sub-symbolic representations" [Russell and Norvig, 2003], prevent both easily explaining the system's results or acquiring a deep understanding of the problem of plan recognition. Thus, this book will not focus on work in this area, and we refer the interested reader to the above citations and the excellent overview of some of this work by Albrecht and Stone [2018].

There are also settings where recognition is relevant, but for which the systems we describe here must be extended. Among the topics that are beyond the scope of this book are multiagent recognition, where the objective is to interpret the behavior of a group of agents, settings where recognition is performed by agents that are active in the system and can adapt their behavior based on recognition of other agents' goals [Levine and Williams, 2014], and settings where the actor reasons about its objectives and may change them during execution. We will have more to say about these issues in Chapter 5 when we consider open research questions.

Instead, this book follows in the footsteps of the early work on what we will call *symbolic plan recognition*. That is, this book focuses on attempts to formalize the knowledge and inference that are needed to perform this specific task. In doing so we believe it is possible to develop efficient algorithms that leverage the specific features of this problem. As such, our focus also rules out the use of general AI or cognitive architectures like SOAR [Laird, 2012] or ACT-R [Anderson, 1976] to recognize plans. This will leave us space in this book to focus specifically on symbolic plan and goal recognition algorithms.

One very abstract formulation of the plan recognition problem that we will focus on in this book is the following.

The Plan Recognition Task

Inputs: A sequence of observed actions O and a set of possible goals, $\mathcal{G}$.

Output: A plan p_g for achieving (not necessarily optimally) a goal, $g \in \mathcal{G}$, that *best satisfies* all the observations in O. We will notate this as $p_g \models O$.

Of course this immediately calls for us to define what we mean by the plan that "best satisfies" the observations. As we will see, even this question does not have just a single answer in the plan recognition community. In fact, this is one of the primary questions that differentiates two of the largest approaches to plan recognition. For example, much of the work we will discuss is actually probabilistic and so one can imagine defining "satisfaction" in terms of a probability distribution over a set of plans that achieve a set of goals.

1.3 SELECTED HISTORICAL PLAN RECOGNITION SYSTEMS

"Those who cannot remember the past are condemned to repeat it." - George Santayana, 1905

In the following subsections we will provide a very brief overview of some significant pieces of prior work in symbolic plan and goal recognition. Our intent here is NOT to provide a compressive list of all the valuable work that has been done in the past, or even to enumerate a complete and covering set of the methods that have been proposed to solve the problems. Instead, these subsections provide a very quick overview of some of the pieces of work that have shaped the overarching trajectory of work in this area. Often the lessons learned by the researchers shaped the research done by following work. Some of these lessons were clearly stated at the time and in other cases they were simply assumed by later researchers and then potentially lost by still later researchers. We will attempt to highlight these points, and we also hope that this set of past works might provide a place for others to start from for their own more comprehensive literature reviews. Note that we will include citations for each, but especially in the case of some of the older work there are a number of papers published about each piece of work and interested readers are encouraged to look further than the citations given below.

1.3.1   KAUTZ AND ALLEN: GENERALIZED PLAN RECOGNITION

Citation: H. Kautz and J. Allen, Generalized plan recognition, *Proceedings of the National Conference on Artificial Intelligence*, pp. 32–38, 1986.[1]
Application Domain: Cooking.
Technical Approach: Graph covering based on a preexisting plan library.
Fundamental Contribution: Plan libraries and problem formalization.
Constraints: Assumed that a minimal graph covering was the best answer.

As we have already pointed out, Schmidt et al. highlighted and began the process of formalizing the problem of plan recognition in 1978, however one of the most cited early works in plan recognition is that of Kautz and Allen almost ten years later. In their work, they formalized the problem of plan recognition, not as that of arbitrary reasoning over a logic describing action, but as a graph covering problem. The plans in the domain of interest (in the case of their work, cooking pasta[2]) were encoded into a *plan library* in a specialized knowledge graph (see Figure 1.1). The plan library encoded:

- a distinguished set of top-level plan classes that were potentially recognizable;

- specialization relations between classes of plans (i.e., making spaghetti marinara is one way of making a pasta dish);

- decomposition relations between plans steps (i.e., in order to make spaghetti marinara you have to make both spaghetti and marinara sauce);

---

[1]Kautz and Allen [1986]
[2]Their work using a cooking as a domains also set the stage for much later work in plan recognition in this same area.

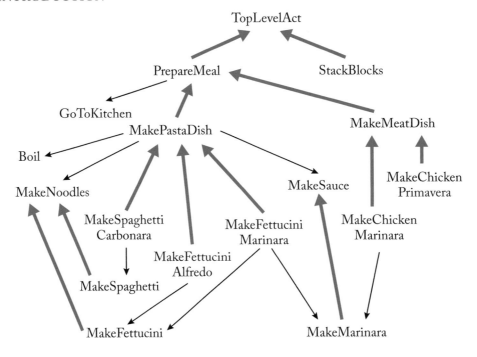

Figure 1.1: Plan graph for making a meal from Kautz and Allen [1986].

- all of the plan steps that needed to be taken in order to successfully execute the plan; as well as

- all necessary ordering constraints between the plan steps.

It is worth explicitly noting that such a representation meant these plans were *hierarchical* or *decompositional* in flavors like those used in some of the early planning systems [Sacerdoti, 1975, Tate, 1977]. This contrasts with non-hierarchical action representations used in other early planning systems like STRIPS [Fikes and Nilsson, 1971] and for work in *plan recognition as planning* (PRaP) that we will discuss later. This use of a plan library to define the set of plans to be recognized and that was explicit and differentiated from other knowledge used by systems is one of the major contributions of this work and sets the tone for much of the following work.

In this system, the recognition of plans was done by presenting an ordered set of observations to the system and using a logic based, minimal graph covering algorithm to find the smallest set of plans that were consistent with the observed actions. As such, it had a number of attractive properties including the ability to recognize multiple plans and that an individual observed action could play a role in multiple plans. Note this was not a probabilistic system, and the inference algorithm followed Occam's razor such that the minimal explanation of the observations should be thought the best. This lead to one of the system's few serious limitations.

It is relatively easy to find domains where a single plan that covers all of the observations is much less likely than an explanation that has multiple concurrent plans being executed, however the work of Kautz and Allen would prefer the single plan result as being more "minimal." For example, if we observe a farmer buying fuel oil and ammonium nitrate it is entirely possible that both purchase events contribute to a single plan to build a "fertilizer bomb" however baring any other information it is far more likely that the farmer just needs these supplies for his farm and each is part of two different but concurrent plans for farming which would violate the minimum covering criteria. That said, this work also made significant contributions to efficient logical inference and knowledge representation.

1.3.2   CARBERRY: PLAN RECOGNITION IN NATURAL LANGUAGE DIALOGUE

Citation: S. Carberry, *Plan Recognition in Natural Language Dialogue*, ACL-MIT Press Series in Natural Language Processing. MIT Press, 1990.[3]
Application Domain: Natural Language Understanding, Discourse.
Technical Approach: Logical inference using the situation calculus.
Fundamental Contribution: Formalizing the vast inference in language.
Constraints: Cost of encoding and inference.

This subsection, unlike any of the others, is not actually about the individual paper cited above. While it is an excellent paper, we have included it because it is one particularly good citation in an entire field of research, namely the use of plan recognition in discourse reasoning and natural language processing.

As we have already alluded to in the Introduction, plan recognition is intimately intertwined in the process of natural language processing and discourse reasoning. Long before the advent of voice-based personal virtual assistants like Apple's Siri and Amazon's ALEXA, researchers in natural language processing recognized that moving beyond processing simple single sentences requires modeling and reasoning about the content of all of preceding sentences in the dialogue and constructing what are called discourse models. Such reasoning crucially involves plan recognition, first in the understanding of the meaning of an individual sentence. Consider again the musical example in the Introduction. Suppose while talking to my friend instead of asking "What song is playing?" we were to be more polite and ask "Do you know what song is playing?" This sentence is a "yes or no" question, but if our friend were to say only "Yes" without providing the additional information we would be right in seeing them as being difficult if not hostile. That said, while such questions are often indirect requests for the specific information, this is not always true. If a software project manager asks one of their subordinates "Do you know how the A* search algorithm works?" they are most likely not interested in an extended discussion of the algorithm, but rather want to know if the subordinate could imple-

---

[3]Carberry [1990]

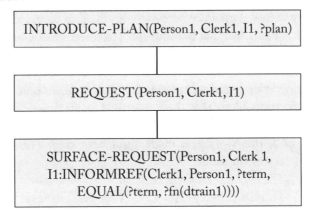

Figure 1.2: Inference of discourse level plan from Litman and Allen [1987].

ment it. In this case, the correct answer would be either "yes" or "no." Correctly answering the question requires inferring the speaker's plan for the provided knowledge and answering accordingly. However, this is far from the only use of plan recognition in natural language processing. As the initial example showed, people frequently answer questions with questions. The only way to make sense of such answers is to correctly infer the plans and goals of the agent.

Much of the early work on plan recognition for natural language processing and discourse reasoning was focused on identifying and formalizing the reasoning to be done. The necessary knowledge for this inference was then encoded in a formal logic, often in extensions to McCarthy's situation calculus [McCarthy and Hayes, 1969] and reasoning was done using specialized theorem provers often in PROLOG. Thus, rather than using a separate and explicit plan library, plan recognition in this work has often encoded all of the same information and relations within the discourse logic.

While critical to formalizing and understanding the reasoning required for natural language processing, this work required the encoding of a great deal of knowledge. Each separate, identified case of recognition was encoded and the logics extended if required. This had the knock-on effect of increasing the complexity and runtime of the logical inference process used and made maintaining such systems difficult and time consuming. While much of the work in discourse has moved on from this approach, it is a large body of research that is full of insights into plan recognition necessary for natural language processing [Allen and Perrault, 1980, Grosz, 1978, Pollack, 1992]. Further, natural language processing is one of the most promising and crucial application areas for the future of plan recognition.

1.3.3 VILAIN: GETTING SERIOUS ABOUT PARSING PLANS: A GRAMMATICAL ANALYSIS OF PLAN RECOGNITION

Citation: M. Vilain, Getting serious about parsing plans: A grammatical analysis of plan recognition, *Proceedings of AAAI-90*, pp. 190–197, 1990.[4]
Application Domain: (None) Complexity analysis
Technical Approach: Complexity analysis of parsing formal plan grammars.
Fundamental Contribution: Key formal results in complexity of the problem and the need for generativity in the problem.
Constraints: No system provided.

Vilain's foundational work on the complexity of plan recognition should be required reading for any researcher in the field. While, as a theoretical analysis this work did not include any discussion of specific algorithms or systems, by drawing connections between hierarchical plan libraries and context-free grammars he was able to establish the first and some of the most important complexity results for the problem of plan recognition. We would call your attention to two specific results in the paper. First, he established that "Recognizing plans with abstraction and partial step order is NP-complete,..." [Vilain, 1990]. While this does establish that plan recognition is a non-trivial problem, notice that both abstraction and partial ordering are necessary for this complexity result to hold. Simpler formulations of this do not have this complexity. For example, simple parsing of context-free grammars without partial ordering are know to have a cubic worst case runtime. In fact, he went further and explicitly pointed out that "An acyclic hierarchy does not contain any recursive plan definitions, and could in fact be encoded as a regular (finite-state) grammar" [Vilain, 1990]. Such plan libraries lack generatively and encode only a finite number of plans. Depending on the context, they are actually addressable by very efficient recognition algorithms like KMP string matching [Knuth et al., 1977]. Thus, it is worthwhile to consider how hard a domain and instance of plan recognition one is dealing with.

1.3.4 GOLDMAN AND CHARNIAK: PROBABILISTIC TEXT UNDERSTANDING

Citation: R. Goldman and E. Charniak, Probabilistic text understanding, *Statistics and Computing*, 2, pp. 105–114, 1992.[5]
Application Domain: Story understanding.
Technical Approach: Dynamically assembly of Bayes nets.
Fundamental Contribution: Plan recognition as probabilistic abduction.
Constraints: Limited by Bayesian methods of the time, and building Bayes nets dynamically.

---

[4]Vilain [1990]
[5]Goldman and Charniak [1992]

This is the earliest work we know of that used Bayes nets for plan recognition. It focused on performing natural language processing on simplified sentences to achieve story understanding. In this case story understanding was viewed as probabilistically answering queries about the story.

Generally, to use Bayes nets for query answering, one would build a single network that covered the entire space of possible situations and queries. Then, evidence (in this case provided by the story) would be presented to the network, and in the face of this evidence the network would be evaluated to compute probabilistic answers to queries. However, because the space of potential stories is effectively infinite, in this case a single Bayes net cannot be constructed *a priori*. Instead, using a set of predefined component Bayes nets that captured the semantics of simple sentences, this work dynamically assembled a larger Bayes network that would cover enough of the space to answer questions about the story. Developing a method to dynamically assemble a Bayes net for plan recognition was a contribution in its own right, but this work also directly addressed the limitation of the Kautz and Allen work by explicitly probabilistically considering cases where multiple plans were happening at the same time.

From a plan recognition perspective, story understanding also presents a set of unique challenges. Most significantly, story understanding often requires the hypothesizing of other, as yet unnamed, individuals. Consider the following very simple story: "Jill saw the 'open' sign on the small corner bakery door, and walked in." Traditional plan recognition should allow us to answer a question like "Why did Jill enter the bakery?" because we recognize the action of entering an open store as a step in buying something there. However, to answer a question like "Did Jill talk to anyone in the bakery?" we need to know a great deal more about the plan for buying. For example, our recognition system would need to know that in the case of small artisanal bakeries that a clerk if not the baker themselves is often present in the shop, and that purchasing goods from them would require talking to them. Thus, correctly answering such questions requires much richer models of plans and reasoning and quantifying over other agents and their roles in the plans being followed.

As such, plan recognition for story understanding is a very interesting and rich application area in its own right. It has significant potential for automated construction of news digests and summarization. Further, this work is significant since, Bayes nets have gone on to be used in a number of other application areas as we will see later in this section.

1.3.5   PYNADATH AND WELLMAN: ACCOUNTING FOR CONTEXT IN PLAN RECOGNITION WITH APPLICATION TO TRAFFIC MONITORING

Citation: D. Pynadath and M. Wellman, Accounting for context in plan recognition with application to traffic monitoring, *Proceedings of UAI-95*, pp. 472–481, 1995.[6]
Application Domain: Driving/lane Selection.

---

[6]Pynadath and Wellman [1995]

Technical Approach: Probabilistic State Dependent Grammar (PSDG) to specialized probabilistic inference.

Fundamental Contribution: Formalizing the problem in a grammar that considered state.

Constraints: Used specialized probabilistic algorithm.

This work presents yet another interesting application area and solution method for the plan recognition problem. At its core this work was concerned with including that state of an actions execution in the probabilistic inference of a plan. Consider a simple sequence of observations. We see someone walk down the street. They see a building on fire with no firefighters in evidence. They pull out their cellphone and place a call. Our first conclusion about their plan is that they are calling the fire department to report the fire. However, if there were firefighters present our first conclusion might be that they were calling a friend to tell them about the fire. Thus, the context in which the actions are observed are crucial to recognizing the correct plan and its next steps.

Well before modern self-driving cars, Pynadath and Wellman used plan recognition to recognize drivers plans and predict lane changes. They did this by formalizing the plans for lane changing in what they called a probabilistic state dependent grammar (PSDG). The PSDG allowed for contextualizing context free grammar production rules on the state of the world and not just on the non-terminal that needed to be expanded. Thus, the grammar production rule for passing car was only appropriate when there was an open lane for passing and sufficient space (see Figure 1.3). Probabilities allowed the system to deal with some noise in the observations. While in this work, the plan recognition problem was formalized in terms of a grammar, the paper itself discusses attempting to implement the inference in terms of a Dynamic Bayes network. This is the first reported use of Dynamic Bayes nets for plan recognition, however the paper reports abandoning this approach in favor of a specialized probabilistic inference algorithm that they developed. Thus, this paper represents both a novel application domain, as well as one negative result for a potential algorithm.

1.3.6 AVRAHAMI-ZILBERBRAND AND KAMINKA: FAST AND COMPLETE SYMBOLIC PLAN RECOGNITION

Citation: D. Avrahami-Zilberbrand and G. Kaminka, Fast and complete symbolic plan recognition, *Proceedings of IJCAI-05*, pp. 653–658, 2005.[7]

Application Domain: RoboCup.

Technical Approach: Marker passing over packed parse trees.

Fundamental Contribution: Very efficient plan recognition.

Constraints: Multiple instances of the same plan and required total ordering.

---

[7]Avrahami-Zilberbrand and Kaminka [2005]

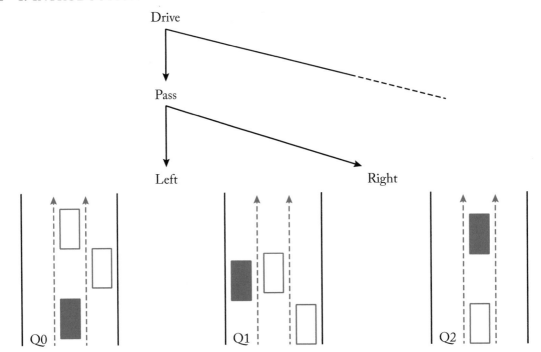

Figure 1.3: Example of a simple PSDG for recognizing driving behavior from Pynadath and Wellman [1995].

This work highlights another approach to plan recognition. In this case, the set of all possible plans represented in a hierarchical plan library (similar to that used in Kautz and Allen's work) is represented using a data structure that is very similar to the "packed parse trees" [Younger, 1967] used for parsing natural language grammars (see Figure 1.4). Once the set of all possible plan structures have been enumerated in this way a simple marker passing algorithm can be used to identify those plans that are consistent with a given set of observed actions. As this was not a probabilistic system, all of the plans consistent with the observations were considered to be equally likely and presented to the user. The use of marker passing in the packed chart proved very efficient.

The most significant limitation of this work was that as a result of its packed chart representation, it was unable to consider multiple concurrent plans within a single hypothesis and plans were assumed to be totally ordered. The efficiency of the representation was predicated on the presence of only one active totally ordered plan in the observation stream.

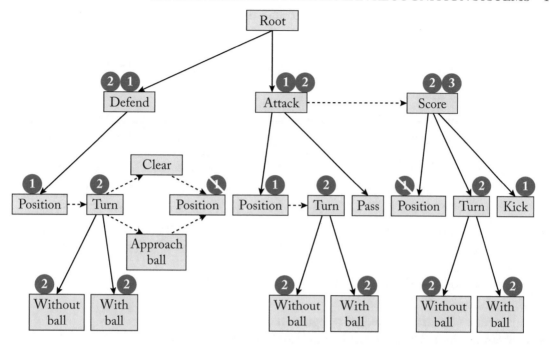

Figure 1.4: Example compressed marker passing plan structures similar to "packed parse trees" from Avrahami-Zilberbrand and Kaminka [2005].

1.3.7   GEIB AND GOLDMAN: A PROBABILISTIC PLAN RECOGNITION ALGORITHM BASED ON PLAN TREE GRAMMARS

Citation: C. Geib and R. Goldman, A probabilistic plan recognition algorithm based on plan tree grammars, *Artificial Intelligence*, 173(11), pp. 1101–1132, 2009.[8]

Application Domain: Synthetic domains.

Technical Approach: Treat plan recognition as parsing of formal grammars.

Fundamental Contribution: Efficient grammars for parsing, multiple concurrent goals, pending sets.

Constraints: Required early commitment to goals and building the complete set of parses.

This was the first work to explicitly advocate viewing plan recognition as that of explicitly parsing a probabilistic grammar representing the plans. Much like the work of Vilain, this work represented the library of plans to be recognized in a formal grammar. However, unlike the work of Pynadath and Wellman or that of Avrahami-Zilberbrand and Kaminka the grammar was not precompiled into another fixed form to aid computation. As such, this system lead to a number of systems working in the area of "plan recognition as parsing." We include it here for

[8]Geib [2009]

this role in the development of the field, but refer the reader to Section 3.2 for a discussion of its functioning.

1.4   SELECTED HISTORICAL GOAL RECOGNITION SYSTEMS

The following subsections will cover some notable goal recognition systems. While goal recognition and plan recognition are different, these systems are significant within the larger research area either for the approach taken or the domain of application.

1.4.1   THE LUMIERE PROJECT: BAYESIAN USER MODELING FOR INFERRING THE GOALS AND NEEDS OF SOFTWARE USERS

Citation: E. Horvitz, J. Breese, D. Heckerman, D. Hovel, and K. Rommelse, The lumiere project: Bayesian user modeling for inferring the goals and needs of software users, *Proceedings of UAI-98*, pp. 256–265, 1998.[9]

Application Domain: Software assistive systems.

Technical Approach: Bayes nets.

Fundamental Contribution: Goal recognition in a real system.

Constraints: Limited by Bayesian methods and computers of the time.

The Lumiere projects out of Microsoft research focused on providing an intelligent help facility for users of Microsoft software. It used Bayes nets to attempt to identify a user's high-level goals and when they were having difficulty so that they could be directed to help resources for using the software. The project was the technology behind the "Clippy the paperclip" help system that was deployed in Microsoft Office for Windows (versions 97 to 2003) after which it was removed because of poor user response.

Despite its eventual removal, this was still one of the pivotal moments in research in this area. First, this work resulted in one of the first very large-scale plan/goal recognition systems deployed in a commercial software environment. Second, as a research prototype it was exceptionally successful and had very high accuracy and recall in goal identification in largely unconstrained domains. Its failure in the Microsoft software ecosystem had much more to do with simplifications of the system that were required to limit the load on then current hardware and the ways in which the recognized plans were use (e.g., users didn't like being interrupted by Clippy.)

---

[9]Horvitz et al. [1998]

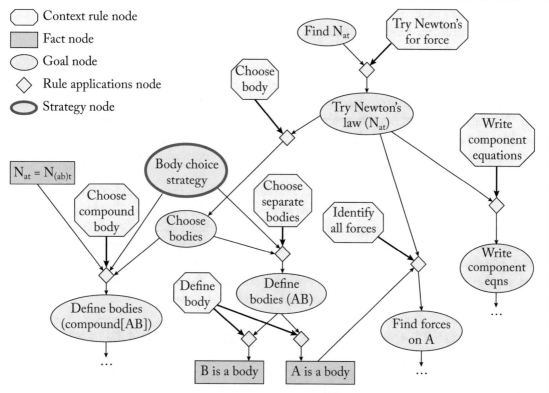

Figure 1.5: Bayesian graphical model for student modeling from Conati et al. [1997].

1.4.2   ON-LINE STUDENT MODELING FOR COACHED PROBLEM SOLVING USING BAYESIAN NETWORKS

Citation: C. Conati, A. Gertner, K. VanLehn, and M. Druzdzel, On-line student modeling for coached problem solving using Bayesian networks, *Proceedings of User Modeling-97*, pp. 231–242, 1997.[10]

Application Domain: Educational agents.

Technical Approach: Bayes nets.

Fundamental Contribution: Explicit modeling of incorrect plans.

Constraints: Bayesian models: propositional and scale.

We would note that this system is not strictly a plan or goal recognition system. However, we have included this work because of its application area. It is one of the most extensive early research projects on intelligent tutoring systems. The Andes system was designed to teach students Newtonian physics and to do knowledge assessment of their abilities. To that end the system

[10]Conati et al. [1997]

watched students in learning skills and attempts to recognize when they have correctly execute the plans for problem solving. When they have gone wrong Andes would provide them with feedback to correctly execute the process the next time.

In order to effectively aid students that are attempting to learn new skills such systems must not only recognize the correct execution of the task to be learned but also those common errors that students make when learning. As a result, such systems must crucially include in their plan libraries and recognize the results of executing common *unsuccessful* and *incorrect* plans. Just like human teachers this allows the Andes system to recognize these common errors of understanding and mistakes and allows the system to take remedial action to clarify the concepts for the human students.

To do its reasoning, the Andes system explicitly modeled the knowledge of the student in a Bayes net. Like other such systems it was therefore limited to a propositional representation, requiring careful knowledge engineering in building the system building, and making it difficult to scale or migrate. However, in the case of this kind of specialized application, this is not as great a limitation as some other applications. Since the system is designed to teach one skill and another instance can be built for a different skill, the knowledge engineering and system building costs can be paid once and amortized over all future uses.

1.4.3 RAMÍREZ AND GEFFNER: PLAN RECOGNITION AS PLANNING

Citation: M. Ramírez and H. Geffner, Plan recognition as planning, *Proceedings of IJCAI*, pp. 1778–1783, 2009.[11]
Application Domain: International Planning Competition domains.
Technical Approach: Computing the distance between plans for known goals and the observed action sequence.
Fundamental Contribution: Viewing the problem as an application of planning technology.
Constraints: Complexity of planning for multiple goals.

This work is the first in a very different school of thought within the plan recognition community. The central insight in this work is that plan recognition problems can be solved by repeated application of AI *planning* algorithms. This has also lead to a sizable number of systems that work in the area of "plan recognition as planning." We include it briefly here for this role in the development of the field, but refer the interested reader to Section 3.3 for a discussion of its functioning.

---

[11]Ramírez and Geffner [2009]

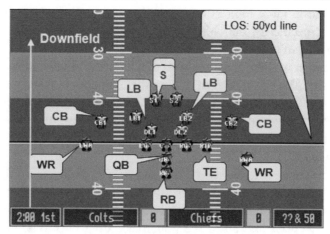

Figure 1.6: Typical game screen from RUSH 2008 football video game from Laviers et al. [2009].

1.5   SELECTED HISTORICAL ACTIVITY RECOGNITION SYSTEMS

In the following we will briefly mention some significant works in activity recognition. We do this both because these works have helped to shape much of the work in plan recognition and also because as we have suggested in the introduction, activity recognition systems can often be used to produce the inputs to plan recognition systems. Thus, understanding the kinds of inputs that are available can be helpful in building plan recognition systems.

1.5.1   IMPROVING OFFENSIVE PERFORMANCE THROUGH OPPONENT MODELING

Citation: K. Laviers, G. Sukthankar, D. Aha, M. Molineaux, and C. Darken, Improving offensive performance through opponent modeling, *Proceedings of AIIDE*, pp. 58–63, 2009.[12]
Application Domain: Video games (RUSH 2008 football).
Technical Approach: Support vector machines for classification.
Fundamental Contribution: Real-time recognition.
Constraints: Learning new plays in a discriminative model.

We include this work for three reasons. First, its domain of application is in the exciting application area of video games. Second, it is an innovative application of support vector machines to the problem of activity recognition. Third, to the best of our knowledge, it is the first such deployed system that worked fast enough to be effective in real time. In this case, activity recog-

[12]Laviers et al. [2009]

nition was used to create more interesting synthetic opponents for human players of RUSH 2008 video football. Specifically, the objective of the work was to infer the defensive play being executed by the human player (red players in Figure 1.6). The results of this inference were then passed to a decision making component so that an appropriate choice can be made about switching the offensive play being used by the synthetic opponent (blue players in Figure 1.6) to make them more challenging players. The core challenge therefore was to recognize the play being executed rapidly enough to allow for changing the offensive play. The system was in fact able to recognize the plays of human users quickly enough that the resulting system was actually a more difficult opponent.

The clearest problem with the use of support vector machines for this kind of work is generalization and the addition of new plays. The addition of a new play requires collecting sufficient training data and retraining the SVNs. In this case, 40 examples from all of the possible offensive and defensive plays and all of the potential stating positions. While this might be problematic, it is not a huge price to pay for rapid activity recognition if it does not require frequent rebuilding or updating.

1.5.2 HIERARCHICAL CONDITIONAL RANDOM FIELDS FOR GPS-BASED ACTIVITY RECOGNITION

Citation: L. Liao, D. Fox, and H. Kautz, Hierarchical conditional random fields for GPS-based activity recognition, *Proceedings of the 12th International Symposium of Robotics Research (ISRR)*, 2005.[13]
Application Domain: Daily activity tracking (2D tracking).
Technical Approach: Hierarchical conditional random fields.
Fundamental Contribution: Activity and location recognition based on real GPS data.
Constraints: Discriminative model based on location.

We highlight this work because, in addition to performing activity recognition on real-world GPS data, it is one of the first works to present a unified algorithm for both activity recognition and significant location identification. This work's goal was to segment a user's day into high-level activities like working, shopping, or travel, and on the basis of this activity model to identify locations associated with these activities, such as workplace, friend's house, or user's bus stop.

The system used a discriminative model (hierarchical conditional random field. It first learns a CRF to spatially segment the GPS data into activities such that the consecutive GPS readings are close to one another (leveraging street map data when available). After this segmented activity model is learned, a second phase of learning is used to construct a second-level CRF is on top of the activity segmentation to label the locations for each activity (see Figure 1.7). This process distinguishes between *navigation activities* (moving from place to place) and *significant activities* (activities done at a particular location).

---

[13]Liao et al. [2005]

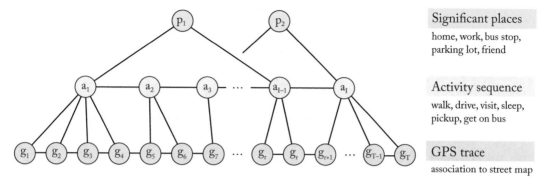

Figure 1.7: Hierarchical conditional random field for behavior recognition from Liao et al. [2005].

As such, it is very impressive work and executed on real world data. That said, like the work on RUSH football discussed above, the use here of discriminative models for segmenting the activities can have the consequence that adding another activity or location to the model can require the retraining of the whole system.

1.5.3   NOTES ON PRIOR WORK IN MARKOV PROCESSES, HIDDEN MARKOV MODELS, AND RELATED WORK

Markov Processes were studied even before 1900 and have been developed into very powerful tools for predicting the unobservable state of a probabilistic system based on observable variables. In a sense, this community was performing activity recognition (and potentially even plan recognition) well before the field was even recognized by the AI community. We refer the interested reader to Russell and Norvig [2003] for an excellent and far more complete introduction to these models however a very brief discussion is worth while at this point.

Consider the HMM in Figure 1.8, that models a simple system of just one unobservable variable X and an observable variable Y. As we can see in the figure, in such models, the variable X at time $t + 1$ only depends on the variable's state at the previous time point. Further, the value of the observable variable Y at time t, depends only on X's value at the same time. Given such a HMM there are two well-known problems *filtering* and *prediction*.

- *Filtering*: is the process of predicting the current hidden state of the model, $P(X_t|Y_{1:t})$, given all of the previous observations.

- *Prediction*: is the process of predicting a future hidden state of the model, $P(X_{t+k}|Y_{1:t}k > 0$, given all of the previous observations.

In our case, the question we can ask is: What if the value of the hidden state was the label for the observed actions? We would then have a model of activity recognition. Or even more

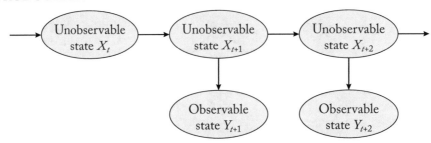

Figure 1.8: A simple HMM model.

advanced, what if the state of the world captured the possible plan states? Then filtering would perfectly capture the problem of plan recognition and even prediction of next actions. This is precisely the realization that motivated the work we will discuss in Chapter 4 on AHMEMs and other Markov Models [Bui et al., 2002, Liao et al., 2005].

There is a truly vast body of literature on HMMs and related Kalman Filters, Conditional Random Fields, and others and we encourage the interested reader to look at it as well. Albrecht and Stone [2018] give an excellent recent review plan and activity recognition work in this area. While far more of it could be seen as falling within the activity recognition camp that the plan recognition camp, it is worthwhile to both give credit where credit is due and to look at and understand this related literature for insights into our problem. At least at an abstract level this work can be seen as formalizing the activity and plan recognition problems even before it was recognized within the AI research community.

1.5.4   FINAL THOUGHTS

Looking back across the collection of systems and approaches discussed here, we would again emphasize, this chapter has covered only a small portion of the exceptional work that has been done on the recognition problem since the field's inception. As such, this review is in no way intended to be a covering or complete review of the work that has been done in the area. Instead it should be seen as a historical arc with jumping off points for researchers to get a handle on the vast number of approaches and domains where work has been done, and give some references to some of the milestones and changes in direction in the field.

With our very rough historical perspective in hand, our next step is to provide a more formal foundation and set of definitions both for recognition problems as a whole and the questions that system designers need to address when thinking about using such systems or developing their own. This is the objective of the next chapter.

CHAPTER 2

Defining a Recognition Problem

While plan, goal, and activity recognition differ in many ways, they share a common structure. In this section we explore this structure and provide a recipe for defining such *recognition problems*. To do this we must bound the scope of our discussion. First, while recognition is necessary to correctly understand behaviors in most multiagent systems, we focus here only on the recognition task of an observer, and not on the potential effects or uses of the conclusions. Second, while in many cases the objective may be to recognize the goals or plans of a set of agents acting, we focus here on the task of interpreting the actions of a single agent.

We describe all recognition problems as having three main components. First, the *environment*, specifying the dynamics of the setting in which recognition occurs. Second, the *actor* representing the agent that is acting in an environment to achieve some objective. We will need to describe how such actors behave within the environment. Third, the *observer* is watching the actor in an effort to recognize its plan or goal. We will have to describe how the actor's actions are perceived and interpreted by the observer. Note that nothing we have said here requires either agent to be human or synthetic. Our formalization here will effectively describe humans recognizing other human's plans, machines recognizing the plans of humans, or any other permutation. In the following subsections we will discuss each of these components in the hope of providing helpful guidelines to the construction of effective recognition systems.

For the remainder of this chapter and book we will use a running example adapted from a synthetic (and simplified) human robot collaboration task by Levine and Williams [2014]. A person named Alice is making breakfast for herself with the help of her trusty robot. We present here just a very intuitive version of the problem here and will provide more formality as it is required later.

Example 2.1   Alice intends to either make coffee (for which she uses a mug) or get some juice (for which she uses a glass). In addition, Alice is having either a bagel with cream cheese or some cereal and milk. Alice's robot assistant is trying to detect her plans and goals, so that it can assist her by, for example, getting the utensils she needs to complete a specific task.

2.1   ENVIRONMENT

The specification of the *environment* describes the dynamics of the setting in which the actor operates, including all aspects of the model that dictate the possible agent actions. It is common to use a compact representation of the environment. Such a representation typically includes the following components. First, a description of the state space S, which capture all the possible states the system may be in. Second, a set of actions A that can be performed by an agent that define state transition functions over S. While the representation formalism for actions may vary, the description of each action includes the domain and range of the state transition function that it defines, and possibly associated probabilities, and time or other costs of execution. Third, in the case of plan and goal recognition, the description typically includes the set of possible goal conditions that implicitly define a set of goal states $G \subseteq S$ agents may try to achieve and the set of possible initial states $I \subseteq S$.

An environment can be represented as a discrete or continuous space. It may be deterministic, when there is a single and known outcome to any action, non-deterministic, if there may be more than one outcome to an action, and stochastic if there is a probability distribution that is associated with action outcomes. Regardless of the representation used, the description of the environment implicitly defines the ways in which the actor can achieve its goals.

When the environment is deterministic, the actor's actions are typically represented by a *plan*, that is a sequence of actions that take the actor from the initial state to some goal. In non-deterministic and stochastic environments, the environment induces a set of policies, which describe the set of actions the actor may perform at each state. Since recognition is typically performed online with the intention of performing goal or plan recognition as early as possible, the recognition analysis is typically performed on the prefixes of plans and policies.

In Example 2.1, a state specifies the position of the objects (e.g., whether the cup is on the table), the execution status of the different sub-tasks, etc. The actions represent the activities that can be performed by the human or the robot (e.g., pour coffee). In our description the outcome of each action is deterministic and known in advance. In a probabilistic version of this problem, an action may fail with some probability, specifying the probability of coffee spilling on the floor when poured.

2.2   ACTOR

Formalizing the **actor** requires enumerating the assumptions made by an observer regarding how an agent with a specific objective will choose to act in an environment.[1] The actor's actions may be influenced by many factors: its familiarity with the environment (possibly reflected by its sensor model), its capabilities and preferences (e.g., can the actor compute an optimal plan?), its

---

[1]Note that these assumptions do not necessarily reflect the actual agent and its capabilities, but in recognition we need to consider how the agent's behavior is interpreted by the observer.

relationship to the observer, and more. Generally, there are three types of relationships between the actor and the observer discussed in the literature.

1. In *keyhole recognition* [Kautz and Allen, 1986], an actor is unaware or indifferent to the recognition of their plans and goals.

2. In *adversarial recognition* the actor is hostile to the inference of its plans and goals.

3. In *intended recognition*, the actor explicitly acts with the intent that its plans and goals are easy to infer.

Returning to our example, if the robot sees Alice picking up the coffee beans, its ability to infer whether Alice intends to make coffee is affected by the assumptions made about Alice and the way she behaves. If Alice is assumed to behave optimally, then picking the coffee beans wouldn't be performed unless Alice intends to make coffee. If, however, the robot knows that Alice maybe trying to deceive it, or is possibly confusing the bag of coffee beans with the bag of oranges, the observation of this action is not enough to guarantee goal recognition.

In addition to the explicit representation of the domain and the agent's relation to the observer, much of the prior work on plan recognition has made use of explicit *plan libraries* [Kautz and Allen, 1986]. Such libraries provide an explicit specification of sequences of actions the actor can execute in the environment to achieve their goals. As such, they define the plans that should be recognized by the system. Historically, most recognition models have relied on such a plan library. Recently, Ramírez and Geffner [2009] suggested that the explicit information in a plan library could be implicitly constructed using AI planning algorithms based on the initial state and the collection of possible goals. As we will discuss in the next chapter, both approaches have their benefits and limitations, and choosing the best one for a particular setting depends on the setting itself and on the objectives of recognition.

2.3   OBSERVER (RECOGNITION SYSTEM)

The actor's representation we explored in the previous section encoded the recognizer's expectations about how the actor behaves with regards to each goal, plan, or activity. We are now shifting focus to the recognizer and the way it performs recognition. We therefore need to describe the recognizer's objectives, the way the actor's actions are perceived and analyzed by the recognizer, and the possible ways that are available for the recognizer to intervene in the recognition process.

2.3.1   RECOGNITION OBJECTIVE

In our recognition settings, we are assuming that the actor enters the environment and follows a policy or plan to achieve its premeditated goal. We consider three possible objectives the recognizer may have.

- **Plan recognition** — where the recognizer's objective is to identify the sequence of actions the actor is in the process of following to achieve its goal.

- ⊖ **Goal recognition** — where the recognizer's objective is to identify the end conditions the actor wishes to a achieve.

- ⊖ **Activity recognition** — where the recognizer's objective is to identify the specific action that is being performed by the actor.

While historically these tasks have often been investigated separately, there is a strong relationship between them, and the distinction between them depends on the application at hand. For example, we have defined the task of making cereal as an atomic activity. However, in a different application, cereal making might be considered a compound action that is comprised of other low-level actions, such as getting the bowl, getting a spoon, etc.

That said, it should be clear that there is a hierarchical relationship between these recognition tasks. In domains where it is required, activity recognition can be seen as producing the observations that are the input to the higher level tasks of goal and plan recognition.

2.3.2   OBSERVABILITY

When considering the way the actor's actions are perceived by the recognizer we need to account for the fact that observability may be hindered due to partial, noisy, and inaccurate sensing. To support this variety of options, the recognizer's sensor model can be viewed as a mapping from action sequences to observation token sequences that may be emitted when the sequence of actions is executed by the actor. The simplest case is when the recognizer has full observability of the actor's actions. This means that when an actor performs some action, a single and unique token is perceived by the recognizer. In partially observable environments, this mapping may be non-deterministic, if there may be more than one token each action execution may emit, and it can be stochastic if there is a probability distribution associated with each token emission.

For example, let's consider the action of getting the milk. If the recognizer enjoys full observability, this action emits a token that allows the recognizer to know that this action was performed. However, suppose the recognizer can only observe that the actor performed an action of taking something from the refrigerator. In this case, the observation token is non-deterministically associated with both the action of taking the milk or the cream-cheese from the refrigerator. In addition, if there is noise in the model, there may be a probability that is assigned to the emission of a token. This may hinder recognition, and needs to be considered when analyzing the actor's observed actions. Note that even when the recognizer is able to recognize the activity of getting the milk with certainty, this does not obviate the need for plan and goal recognition. The recognized activity may participate in multiple plans and goals. For example, getting milk is required for both making coffee and cereal.

2.3.3   POSSIBLE INTERVENTIONS

In most prior work on recognition, beyond recognizing the actor's goals and plans the observer is assumed to be passive. Recent work discusses cases in which the observer (or an agent acting in its behalf) may have a way to control and modify the recognition setting and consequently the way actors behave to achieve their goals [Bisson et al., 2011, Keren et al., 2014, 2019, Mirsky et al., 2018, Wayllace et al., 2017, 2020]. Different interventions are possible in different applications.

In settings with *online interventions*, the observer has the ability to dynamically change the recognition setting while the actor is executing its plan. Such interventions can support an inter-action of the observer with the environment or the actors. For example, by selectively changing some features of the environment, the observer may be able to provoke the actor to behave in a specific way that reveals its intention [Bisson et al., 2011]. Similarly, Mirsky et al. [2018] support settings in which the observer can iteratively query an actor about some environmental features to ensure early recognition. The decision of which query to pose is based on the likelihood of the different goals that are related to the query and its potential information-gain given the current probabilities of each goal. Shvo and McIlraith [2020a] support an observer that can dynamically decide to sense specific environmental features or act in the environment in a way that expedites the recognition of an actor's goal. The decision of which intervention to perform is based on the analysis of each goal's *landmarks*, facts that hold for all plans that achieve that goal.

In the case of *offline interventions*, the observer can modify the recognition setting before actor arrives. For example, recently Goal Recognition Design [Keren et al., 2014, 2019, Mirsky et al., 2019a, Wayllace et al., 2017] has been suggested to provide tools for using offline design to facilitate goal recognition. Specifically, this design process aims at minimizing the worst case distinctiveness (*wcd*), which represent the maximal number of steps an actor can take before its goal can be recognized.[2]

In our running example, an online intervention may correspond to the robot's ability to ask Alice whether she is planning to have juice or coffee for breakfast. An offline intervention could include placing the coffee beans away from the refrigerator (where the juice is stored) forcing the actor to move away from the refrigerator to get the coffee (when desired) and thus helping the observer to distinguish Alice's plans and goals earlier.

2.4   DESIGN CONSIDERATIONS

Having discussed the basic elements of recognition, we now specify a list of issues that should be characterized in order to define a recognition problem. This list includes some of the questions that should be answered to make sure a recognition problem is well defined.

Environment

✓ What are the dynamics of the environment?

✓ How is the state space represented?

✓ What are the possible actions an agent can perform at each state?

---

[2]See Keren et al. [2020] for a recent survey on Goal Recognition Design.

✓ How are the effects of actions defined? Are actions deterministic, non-deterministic, or stochastic?

Actor

✓ What are the assumptions made by an observer about an actor's actions?

✓ What is the best way to represent patterns of actor actions?

✓ What are the assumptions made about an actor's relationship to the recognizer? Is the actor adversarial, collaborative, or agnostic to the recognizer and the recognition process?

Observer

✓ What is the recognition objective? Is the observer trying to recognize an actor's goal, plan, or activity?

✓ How are the actor's actions perceived by the observer? Does the observer enjoy full observability of an actor's states or performed actions? Otherwise, how is the observer's sensor model defined?

✓ What are the possible interventions an observer can perform to facilitate recognition?

In the next section, we explore multiple approaches to recognition that have answered the above questions in a variety of ways.

CHAPTER 3

Implicit vs. Explicit Representation of Knowledge

For much of the rest of this book we will discuss the specifics of systems, their contributions, and individual strengths. However, plan recognition research is often distinguished by the kind of knowledge available to the system and how it is represented. Before distinguishing these representational choices, we first discuss give some unifying structure and shared assumptions.

3.1 SHARED ASSUMPTIONS

The approaches that we focus on in this book share four assumptions.

1. **Model-based recognition.** The recognizer has a model of the observable actions the actor can perform. While there has been a small amount of work on model-free approaches for goal and plan recognition, usually in the form of neural networks, these systems are not covered here.

2. **Observer perspective recognition.** Recognition is based on the observer's model of the domain. That is, these algorithms don't make claims about the mental model of the actor, but rather the observers' best idea about what the actor is doing and why. This requires the observations of the actor's actions to be mapped to actions within the observer's model. We have already discussed this as activity recognition, but it is worth noting that this mapping can be a source of errors.

3. **Structural recognition.** As we have said, a problem is considered a plan recognition problem only if it constructs a model of the plan being followed beyond the actions already executed by the actor. Such a model can enable the observer to predict which future actions might be taken by the actor.

4. **Prefix recognition.** For most applications, it is crucial that a plan is recognized before it is completed. This enables the observer to take actions to aid in the plan, thwart the plan, or perhaps just slightly modify the state archived by the actor. To do this kind of *incremental* or *online* recognition, algorithms cannot assume they have seen the entire plan. That is, they must work on proper prefixes of a complete plan execution trace. Note, this does not preclude algorithms that also work on complete plan execution traces.

Given these shared assumptions, work in plan recognition is often divided into two major subgroups based on having an *implicit* or *explicit* (see Section 2.2) representation of the plans to be recognized.

Implicit Representation means that we have a representation of an initial state, a goal(s), and actions in the form of state transition functions that describe what the actor can do in the world. Such representations do not contain further information on *how* the actor should achieve a particular goal, and the recognition task usually revolves around reconstructing a valid plan for the goal that includes the observations. Systems using this approach often represent their domain knowledge using classical planning models, MDPs, Partially Observable Markov Decision Processes (POMDPs), and games. Ramírez and Geffner [2009] that popularized the idea of *Plan Recognition as Planning* is a representative model in this family we will discuss later in this chapter.

Explicit Representation means that in addition to a list of symbols representing the actions that can be observed, the domain model also includes specific knowledge of *how* a goal be can achieved, potentially requiring all the valid variations to be explicitly specified. Recognition systems that utilize these representations usually care about the structure and the context of the executed plan. Systems using this approach often represent this domain knowledge using formal grammars, and or trees, plan graphs, and even HMMs. For example, later in this chapter we will discuss *Plan Recognition as Parsing* as advocated by Geib and Goldman [2009b].

We have chosen the terminology of *implicit* and *explicit* to highlight the fact that knowledge about what plans are to be recognized by these systems MUST be defined in some form. The set of plans to be recognized are either explicitly defined in the domain theory or are implicitly defined by the domain theory's interaction with a planning system. While some research doesn't fall neatly into one of these two areas, this division does provide a starting point. When trying to tackle a new problem, asking *"Do we have an explicit or implicit representation of the plans to be recognized?"* can help focus on the type of solutions to apply.

There is also a significant difference in the kinds of knowledge used in these two representations. Because most of the knowledge in an explicit representation is about the decomposition of high level actions into lower level ones and ordering constraints over them, such systems often work in an *action space* rather than a *state space*. That is, they search for a sequence of actions consistent with the domain knowledge that is the prefix of a plan matching the observed actions. They do not search in the causal state space of the plans used in implicit representation recognizers, but in the space of action sequences. Such systems may not even represent the state of the world. Perhaps ironically, explicit representations often leave as implicit the causal knowledge that define individual actions. In domains where building sufficiently complete causal models of the actions is difficult or impossible, this may be an argument in favor of explicit representations of the plans for recognition.

The next two sections will formalize these approaches: explicit—as working in an action space and implicit—working in a state space. We will draw as many parallels as possible between the two approaches. In each case we will start with some introductory remarks, followed by formally defining a domain theory and problem instance within the representation. Each section will then demonstrate the approach in a shared general example for the breakfast problem discussed in Example 2.1. Remember that in this domain Alice is making breakfast with the help of her trusty robot. Alice can make coffee (for which she uses a mug) or get some juice (for which she uses a glass). In addition, Alice is having either a bagel with cream cheese or some cereal and milk. Alice's robot assistant is trying to detect her plans and goals, so that it can assist her by, for example, getting the utensils she needs to complete a specific task. The sections will then end by discussing a specific system that used the representational method and how it would be applied to the specific example domain. As the earliest work in plan recognition used explicit domain representations, we will follow historical precedent and discuss explicitly represented domain theories first.

3.2   EXPLICIT REPRESENTATIONS

Plan recognition systems with explicit plan representations assume that the observer has, usually modular, knowledge of the plans to be recognized. This means that in addition to a set of actions that are available in the domain theory, the agent has explicit knowledge of how those actions can be combined to achieve particular goals. This idea goes back to a distinction within the AI *planning* community about the presence of the same kind of knowledge.

A simple agent that is only able to pick things up, put them down, move around, and push buttons would still be able to execute all the actions necessary for our example. In fact it would be able to do a much more, for example making a buffet breakfast for thirty guests. However, to do this without additional knowledge requires searching a truly vast space of possible sequences of actions to produce its plan. An agent with explicit knowledge of how to go about building such a plan can accomplish this same planning task much faster.

Knowledge that to prepare a buffet requires preparing, beverages, salads, cheeses, and cold cuts, multiple kinds of pastries, eggs, sausage, bacon, and even pancakes, allows an agent to divide the problem into these ordered sub-tasks, and plan for each of them independently. Further our agent could have knowledge about how to decompose the sub-tasks into still lower-level tasks (e.g., making pancakes requires making the batter, heating the griddle, pouring the batter, etc.). Such a decompositional knowledge and planning process can be repeated until the plan is ground out in the agent's executable actions. This kind of knowledge can also be used for plan recognition.

Such systems explicitly capture the agent's knowledge of plans as a set of decomposition rules that break down a high-level task into ordered sets of sub-tasks, and usually make three assumptions.

1. All of the possible components of any decomposition are either executable actions or sub-tasks that can be further decomposed.

2. All observable actions play a role in decomposing at least one higher-level goal.

3. All of the plans to be recognized are captured by some set of decompositions.

This kind of decompositional knowledge is most often captured in a *Plan Library* using a formal grammar or a set of plan graphs. In such plan libraries, the observations of an agent's actions are captured in the leaves of plan graphs or terminal symbols of grammar production rules. The plan recognition task is then to find valid plans that are consistent with both the observed agent's actions and the plan library. We will call such plan structures *explanations*, and it is worth noting that a sequence of observations may have a very large number of consistent explanations. It is then the task of probabilistic plan recognition systems to find the most likely.

Plan libraries often contain recursive plan knowledge (e.g., recursive grammar rules or recursive plan graph fragments) that can capture plans with an unbounded repetition of a set of actions and hence plans of arbitrary length. The set of plans encoded in such a recursive plan library is actually infinite. Therefore, number of possible plans that can be recognized by a complete plan recognition algorithm based on such a grammar is infinite. We would also note, that many plan library models have at least context-free expressiveness.

In addition to using formal grammars to represent plans, some plan recognition systems are heavily informed by natural language parsing (NLP) algorithms. That is, they use lessons learned from NLP algorithms to recognize a sequence of observed action "tokens." However, plan recognition systems face three problems not common in NLP.

- *Recognizing the prefix of the plan.* As we have already argued, recognizing a plan before it is complete is crucial for most applications of plan recognition. However, there are NLP systems that leverage the finite length of a sentence and knowledge of periods and other punctuation to aid in parsing. However, plans can be of arbitrary length and it is rare to be certain that a plan has ended. As such, it is safer for our plan recognition systems to take lessons from *incremental* NLP algorithms. Such algorithms stress the processing of each observed token (word or action) as it is observed, in much the way we are often able to anticipate the ending of other peoples sentences. This is also the technology behind things like Google's auto completion of search terms.

- *Plans that repeat a sequence of actions multiple times.* This would be equivalent to repeating the same words in the same order in a sentence multiple times. While this can happen in NLP, it is rare and requires unusual circumstances. In plan recognition it is very common.

- *Concurrent execution of multiple interleaved plans.* Imagine a speaker trying to say two sentences at the same time, alternating the words of the two sentences. This simply isn't done in any human languages; however, it is common in plan recognition settings. Imagine recognizing the plans of a single agent both cleaning the kitchen and doing the laundry at the same time.

Therefore, we must be careful leveraging NLP algorithms.

There are many ways to represent such decompositional domain knowledge with some representations allowing partial ordering, parameterized actions, conditional and posterior probabilities, and more. In the following we will try to capture as many commonalities as we can while formalizing this problem using ideas from formal grammars. That said, similar definitions can be produced for graph-based domain theories and other representations.

3.2.1 FORMULATING THE PROBLEM

We will define a domain theory as a tuple $\mathcal{T}^{\mathcal{E}} = \langle \mathcal{A}, NT, R \rangle$ where:

- $\mathcal{A}$ is a set of terminal symbols, that capture actions that can be observed. (These will be the same action symbols used in an implicit representations but without capturing causal knowledge. See Section 3.3.1.) They will be denoted in bold face Times font (e.g., $\mathbf{o}_i$).

- NT is a set of non-terminal symbols. Each such symbol captures an *abstract action* or *task* that captures a potentially large set of sequences of actions. These will be denoted in typewriter font (e.g., S, g).

- R is a set of production rules of the form:

$$S \to \sigma_1, \ldots, \sigma_n, \tag{3.1}$$

  where $S \in NT$ is the abstract action the rule decomposes and $\forall \sigma_t \in \{\sigma_1, \ldots \sigma_n\} | \sigma_t \in \{NT \cup \mathcal{A}\}$ are the ordered sub-tasks of the decomposition.

Each rule captures one of the ways a given abstract action can be decomposed into more basic sub-tasks. We will refer to just the rules in the domain theory as the *plan library*. Building or recognizing a plan in this formalism requires using the plan library to rewrite non-terminals into equivalent sequences of terminal and non-terminal symbols.

For example, given a domain theory such that $S \in NT$ and

$$S \to \sigma_k, \ldots, \sigma_m \in R, \tag{3.2}$$

we can replace the symbol S in a sequence of symbols with the right-hand side of the domain theory's rule. This expands the sequence of symbols, potentially introducing new symbols between existing symbols. We will call this process *derivation* and denoted it with, $\mapsto$. For example:

$$\sigma_1, \ldots, \sigma_i, S, \sigma_{i+1}, \ldots, \sigma_n \mapsto \sigma_1, \ldots, \sigma_i, \sigma_k, \ldots, \sigma_m, \sigma_{i+1}, \ldots, \sigma_n. \tag{3.3}$$

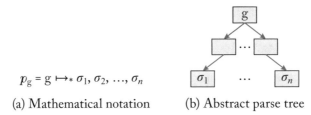

$$p_g = g \longmapsto_* \sigma_1, \sigma_2, \ldots, \sigma_n$$

(a) Mathematical notation   (b) Abstract parse tree

Figure 3.1: Two different representations of the same derivation.

Note that the above assumes that $\forall \sigma_j, \sigma_j \in \{NT \cup \mathcal{A}\}$. We will use $\longmapsto_*$ to denote a derivation of zero or more steps.

Definition 3.1   Given a domain theory $\mathcal{T}^{\mathcal{E}} = \langle \mathcal{A}, NT, R \rangle$, and a goal nonterminal, $g \in NT$, a *hierarchical plan*, p_g, is a tree data structure that captures a set of derivations starting with just the symbol g.

For example, Figure 3.1a denotes the fact that a given plan, p_g, for the goal, g, is a derivation stating with just g. Figure 3.1b captures an abstract tree representation of the same parse of the p_g from g where each level of the tree would be the application of one of the production rules in the plan library.

Definition 3.2   We will call the in-order sequence of symbols on the leaf nodes of a given plan, p_g, the *frontier* of the plan, and denote it $F(p_g)$:

$$F(p_g) = [\sigma_1, \sigma_2, \ldots, \sigma_n] | \sigma_i \in \{NT \cup \mathcal{A}\}.$$

Definition 3.3   We define a plan as *complete* if its frontier only contains terminal symbols. $F(p_g) = [\sigma_1, \sigma_2, \ldots, \sigma_n]$ and $\forall \sigma_i \in F(p_g) | \sigma_i \in \mathcal{A}$.

Note that hierarchical plans are not required to be complete.

Because with explicit representations we define a plan on the basis of derivations and a given domain theory can have multiple production rules that share the same left-hand side symbol, a given non-terminal can potentially have a very large and diverse yield. Further, the same set of actions might be the frontier of plans for multiple different goals. Figure 3.2 shows an example of such a scenario. Observing the actions [a, b, c] might be explained by the upper plan tree, as sub-actions to achieve Y, but might also be explained as sub-actions to achieve T.

Definition 3.4   Given a domain theory, $\mathcal{T}^{\mathcal{E}} = \langle \mathcal{A}, NT, R \rangle$, we define an *instance*, i, of an explicitly defined recognition problem as a tuple, $i = \langle \mathcal{T}^{\mathcal{E}}, \mathcal{G}, O \rangle$, such that:

- $\mathcal{G} \subseteq NT$, is a set of abstract actions or tasks the actor might have as their goals. Notice that unlike an implicit domain theory (see Section 3.3.2) the possible goals are abstract actions that defines a set of possible sequences of actions. We will denote a single goal as g$\in \mathcal{G}$.

- $O = [\mathbf{o}_1, \ldots, \mathbf{o}_k]$ such that $\mathbf{o}_i \in \mathcal{A}$ is an ordered sequence of observed actions the agent executed.

Definition 3.5 Finally, we will define a plan, p_g with a frontier $F(p_g) = [\sigma_1, \sigma_2, \ldots, \sigma_n]$ as *satisfying* an observation sequence $O = [\mathbf{o}_1, \ldots, \mathbf{o}_k]$ such that $\mathbf{o}_i \in \mathcal{A}$ if:

1. $k \leq n$.

2. There is a monotonic mapping of actions from O to $F(p_g)$, such that for each observation $\mathbf{o}_i$ there is a terminal leaf node σ_i labeled with $\mathbf{o}_i$.

We will denote this as $p_g \vDash_{\varepsilon} O$.

The Plan Recognition Task

So, given an instance of the plan recognition task, $i = \langle \mathcal{T}^{\varepsilon}, \mathcal{G}, O \rangle$, the plan recognizer's task is to find a complete hierarchical plan, p_g for achieving some goal g$\in \mathcal{G}$, that satisfies O. This is notated by $p_g \vDash_{\varepsilon} O$.

Notice that (like the definition we provide in the implicit representation case) this definition allows for the plan to be partial and for the mapping to O to be a partial plan rather than a complete plan. Also (like the implicit case), the goal recognition task is a byproduct of the plan recognition task. The goal of the actor is simply the root node of the hierarchical plan. Many of the algorithms that use this approach are probabilistic, and hence may produce a distribution over a number of different hierarchical plans.

3.2.2 RUNNING EXAMPLE

One possible $\mathcal{T}^{\varepsilon} = \langle \mathcal{A}, NT, R \rangle$, for the breakfast example, is:

- $\mathcal{A}$ = { **GetGlass(****), JuiceOranges(** **) , GetMug(** **), GrindBeans(** **), GetBagel(** **), SpreadCreamCheese(** **), GetBowl(** **), PourCereal, DrinkCoffee, DrinkJuice(** **), EatBagel, EatCereal}**

  Note the same actions are used in the implicit example (see Section 3.3.2). In order to slightly shorten some of the sequences of observed actions that we will discuss later, for

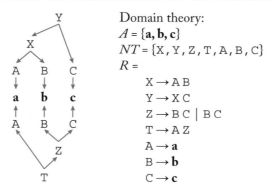

Domain theory:
$A = \{\mathbf{a}, \mathbf{b}, \mathbf{c}\}$
$NT = \{X, Y, Z, T, A, B, C\}$
$R =$

$$X \rightarrow A\,B$$
$$Y \rightarrow X\,C$$
$$Z \rightarrow B\,C \mid B\,C$$
$$T \rightarrow A\,Z$$
$$A \rightarrow \mathbf{a}$$
$$B \rightarrow \mathbf{b}$$
$$C \rightarrow \mathbf{c}$$

Figure 3.2: Example domain theory with plan library and two different derivations for the same set of observations.

those we will use frequently, we have also provided an icon to represent each observable action and where it will shorten description without sacrificing clarity we will use them instead.

- $NT = \{$ MakeBreakfast, MakeDrink, MakeFood, MakeCereal, MakeBagel, MakeCoffee, MakeJuice, GetGlass, GetMug, JuiceOranges, GrindCoffee, GetBagel, SpreadCreamCheese, GetBowl, PourCereal, DrinkCoffee, DrinkJuice, EatBagel, EatCereal$\}$

- $R = \{$

    MakeBreakfast→ MakeDrink, MakeFood | MakeFood, MakeDrink.

    MakeDrink→ MakeJuice | MakeCoffee.

    MakeFood→ MakeBagel | MakeCereal.

    MakeJuice→ GetGlass, JuiceOranges.

    MakeCoffee→ GetMug, GrindCoffee | GrindCoffee, GetMug.

    MakeBagel→ GetBagel, SpreadCreamCheese.

    MakeCereal→ GetBowl, PourCereal.

    GetGlass→ **GetGlass(** ⊔ **).**

    GetMug→ **GetMug(** ☕ **).**

    GetBagel→ **GetBagel(** ◉ **).**

    GetBowl→ **GetBowl(** ⌣ **).**

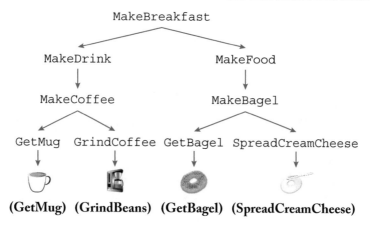

Figure 3.3: Example explanation in the form of a plan tree.

JuiceOranges→ **JuiceOranges(** **)**.

GrindCoffee→ **GrindBeans(****)**.

SpreadCreamCheese→ **SpreadCreamCheese(****)**.

PourCereal→ **PourCereal**.

DrinkJuice→ **DrinkJuice(****)**.

DrinkCoffee→ **DrinkCoffee**.

EatBagel→ **EatBagel**.

EatCereal→ **EatCereal**.

      }

Note the vertical bar in these rules is a short hand for having multiple rules with the same left-hand side non-terminal. For example, the first rule specifies two rules for MakeBreakfast that allow MakeDrink and MakeFood to be done in either order.

Given this plan library, the goal task MakeBreakfast, and the following observation sequence:

[, , ,]

A plan recognizer could produce the hierarchical plan shown in Figure 3.3. Some algorithms will output a set of hypotheses, such that each hypothesis is a set of plan trees. Others will only provide a single explanation. Some algorithms will output a probability distribution over the goal symbols. Some algorithms will not output the tree structure, but just the predictions over future observations, while others will only show the parts of the tree that were already observed. In the next section we will discuss one of the early "plan recognition as parsing" systems that uses an explicit representation.

3.2.3   PLAN RECOGNITION AS PARSING

Citation: C. W. Geib and R. P. Goldman, A probabilistic plan recognition algorithm based on plan tree grammars, *Artificial Intelligence*, 173(11):1101–1132, 2009.[1]
Application Domain: Artificial plan libraries.
Technical Approach: Parsing techniques using probabilistic plan tree grammar.
Fundamental Contribution: First to treat plan recognition as a problem of parsing with ambiguity.
Limitations: High computational costs from of early commitment to root goals computing the complete set of parses and.

The Probabilistic Hostile Agent Task Tracker *(PHATT)* is a plan recognition system based on a model of plan execution. This means that using its grammar and semantics it is able to actively reason about execution of plans for multiple interleaved root goals, efficiently reason about partially ordered plans, and even failing to observe actions.

The PHATT plan library is captured in a probabilistic, context-free, *tree* grammar. Thus, its grammar is specified in terms of a collection of tree structures, not unlike tree adjoining grammars [Joshi and Schabes, 1997]. These tree grammars are made up out of left-most, depth-first trees, and are very much like Greibach normal form rules for context-free grammars [Greibach, 1965]. That is, in each PHATT grammar tree the left-most symbol in the tree's frontier is an observable action. As such, the complete PHATT tree grammar for Example 3.2.2 looks very different than our example (see Figure 3.5). The translation to a left-most tree grammar effectively "precompiles" some of the search required by other systems into the grammar. PHATT tree grammars also support partial ordering, and the system assumes that any of the nonterminals in the grammar could be goals for the agent. This enables the recognition of the execution (and interleaving) of more than one plan at a time.

Based on the state of the world and the previously observed actions, PHATT maintains a *pending set* of those trees that are consistent with the previously observed actions and currently hypothesized partial plans (see Figure 3.4). A generative "ball and urn" probability model based on the pending sets determines the which tree is selected given the next observation. Parsing is performed by performing *tree adjunction* [Joshi and Schabes, 1997] of the newly selected tree to the existing trees in each possible parse. This makes PHATT an incremental parsing algorithm.

The output of the PHATT algorithm is a collection of forests of possibly incomplete hierarchical plans. Each such collection of plans represents a hypothesis or *explanation* about

---

[1]Geib and Goldman [2009b]

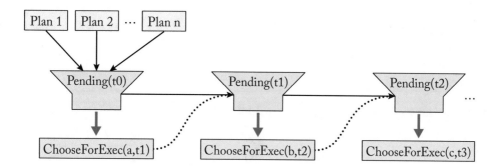

Figure 3.4: Pending sets being progressed from time step to time step from Geib and Goldman [2009a].

the plans being executed and their state. Starting from the left, all of the leaves of the tree are the observed actions, and the remaining leaves are nonterminals of the grammar capturing sub-plans that will be executed in the future. Using the system's pending sets, PHATT's probability model is based on weighted model counting of each of the explanations where the probability of each explanation is the likelihood of the given sequence given the set of all possible enabled plan sequences.

PHATT does have two significant limitations. First, in order to compute accurate prob-abilities, it computes all of the possible parses for a given set of observations. This can be quite computationally expensive, and only short sequences of observations are viable if the plan gram-mar has significant ambiguity. Second, its reliance on left-most depth-first trees means that upon seeing the very first potential action for a plan, the root of the plan must be hypothesized. For example, observing an agent pickup their car keys, PHATT would be forced to consider all of the possible things the agent could do in their car as potential explanations for the observations. It is unable to delay hypothesizing the higher-level goals, and this can be quite computationally costly. We will see this in the following example. However, this was inefficient and motivated much of the work that followed.

Example 3.6   Suppose we consider the full domain theory for PHATT that is equivalent to that in Example 3.2.2 and the example observation sequence:

$$[\;\cup,\;\blacksquare,\;\bullet\;] \tag{3.4}$$

Notice that seeing the same observations in any order is actually a valid observation sequence for PHATT, because GetMug and GrindCoffee are unordered in PHATT's tree grammar, as are MakeDrink and MakeFood.

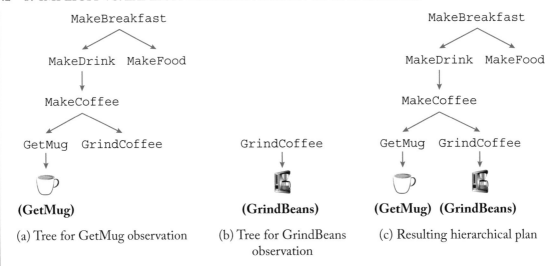

(a) Tree for GetMug observation

(b) Tree for GrindBeans observation

(c) Resulting hierarchical plan

Figure 3.5: Two PHATT grammar trees, (a) and (b), combine to produce hierarchical plan (c) consistent with both observations.

PHATT would first select a left-most tree with the observed action ⎕ and rooted at the symbol MakeBreakfast (see Figure 3.5a). As we said, PHATT's left-most trees force it to consider high level goals even when very little evidence has been seen. This tree would constitute an explanation for just the first observation. When the next observation, ⬛, is processed, a very small left-most tree with the foot ⬛ and the root GrindCoffee (Figure 3.5b) would be combined into the first tree resulting in the tree shown in Figure 3.5c creating a single explanation for the observation sequence.

Given the partial ordering within this domain theory, in addition to this explanation, PHATT would also compute all the other possible explanations for this set of observations. This includes the possibility that each of the actions was done as part of its own plan and each of the possible pairwise groupings of observations into plans. It would then use its probability model to compute the probability of each explanation and the conditional probability of the goals. In the end it would determine that MakeBreakfast was the most likely goal and that the plan shown in Figure 3.5b is in fact the most likely explanation.

3.3   IMPLICIT REPRESENTATIONS

When performing plan recognition using implicit representations, the observer is assumed to have access to causal descriptions of the domain's actions, in the form of precondition and effects rules defined on the environment. Since their first introduction this kind of representation has

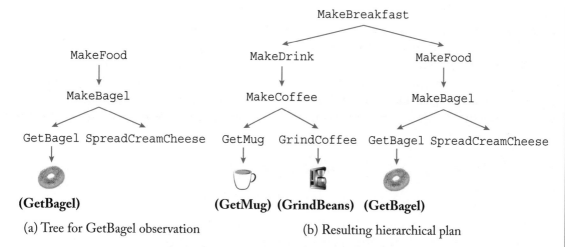

(a) Tree for GetBagel observation   (b) Resulting hierarchical plan

Figure 3.6: A third PHATT grammar tree (a) is combined with the hierarchical plan in Figure 3.5c to produce hierarchical plan (b) consistent with all three observations.

been significantly extended [Ramírez and Geffner, 2010]. In their most general form, such representations may support stochastic actions (i.e., they only probabilistically achieve their effects), may define the goal to be achieved using a reward or utility function rather than a goal condition, and may be only partially observable (i.e., not all of the actor's actions are always observable). While we will discuss some of these extensions in later chapters, this chapter will focus on the most common and simplest implicit representation of the *domain theory* where all the actions are deterministic and assumed to be fully observable.

With the implicit approach, beyond the specification of the domain theory, there is no need to specify how actions can be combined to achieve specific outcomes. Instead, we use a planner in order to generate goal directed behavior and to associate a sequence of observations to a goal. Thus, rather than a specific subset of possible plans being recognized, as in the explicit case, in the implicit case, any plan that can be found by the AI planner can be recognized. Effectively, implicit plan recognizers trade the greater offline and memory costs of representing and storing the plans to be recognized by explicit recognizers, for the potentially smaller offline cost of building causal models of the actions, and the online cost of computing viable plans using modern AI planners.

If we have a set of observations and we have access to a planner that can find an optimal plan for each goal, we can imagine comparing an observed sequence of actions to an optimal plan for each goal. We can then assume that the goal with the closest plan to the observed trajectory would be the most likely. This is the intuition behind much of the work using implicit domain representations. As we will see later on, this basic intuition can be extended to support other forms of planning domains, including continuous actions, stochastic actions, and any representation supported by a planner that an actor may use to plan its behavior.

In plan recognition problems these formulations have often been rooted in automated AI planning research which has the Planning Domain Definition Language (PDDL) [McDermott et al., 1998] to define domain theories. However, because plan recognition research often goes beyond standard PDDL, and we want to establish commonalities between different approaches to plan recognition, rather than using explicit PDDL we will provide more formal syntax for our definitions.

3.3.1   FORMULATING THE PROBLEM

Before we can define a domain theory, some initial terms are required.

Definition 3.7   A set of *fluents*, $\mathcal{F}$, is a set of logical terms. The set of all possible truth value assignments to the elements of $\mathcal{F}$ defines a space of *possible worlds*, $\mathcal{S}_{\mathcal{F}}$.

Definition 3.8   A *state*, $s \in \mathcal{S}_{\mathcal{F}}$, is a complete assignment of truth values to the elements of $\mathcal{F}$.

Definition 3.9   A *condition*, $c \subseteq \mathcal{F}$, is a partial assignment of truth values to the fluents $\mathcal{F}$, and will be denoted with typewritter font (e.g., c, g).

We note that we will be using conditions to define the possible goals of the observed agents. Since they share the same conceptual role, it is appropriate that they share the same typography as goals in the explicit representation (see Section 3.2). However, it is worth highlighting the very real difference between goals as a conditions over fluents in this representation, and goals as a non-terminal symbols in the explicit representation's grammars.

We can now define a domain theory for use in a plan recognition system that uses an implicit representation.

Definition 3.10   For an implicit domain representations, we define a *domain theory* as a pair $\mathcal{T}^{\mathcal{I}} = \langle \mathcal{F}, \mathcal{A} \rangle$, where:

- $\mathcal{F}$ a set of *fluents* that define the set of all possible world states, $\mathcal{S}_{\mathcal{F}}$; and

- $\mathcal{A}$ is a set of actions, such that $\mathbf{a} \in \mathcal{A}$ and $\mathbf{a} = \langle \text{pre}(\mathbf{a}), \text{post}(\mathbf{a}), cost(\mathbf{a}) \rangle$, such that:

  - $\text{pre}(\mathbf{a}) \subseteq \mathcal{F}$ is a condition (called a *precondition*) that specifies when the action can be executed.

  - $\text{post}(\mathbf{a}) \subseteq \mathcal{F}$ is a conditions (called a *postcondition*) that specifies the results of executing the action.

    – $cost(\mathbf{a})$ is the cost of executing the action.

    We will use the same action symbols used in an explicit representations. See Section 3.2.1, and will be denoted in boldface Times font (e.g., $\mathbf{a}_i$).

In the following examples we will denote the pre($\mathbf{a}$) and post($\mathbf{a}$) conditions as sets of possibly negated fluents. In pre($\mathbf{a}$), negated fluents must be false, and non-negated fluents must be true in a state for the action to executable. In contrast, post($\mathbf{a}$) can be used to update the state of the world. Fluents that are negated in post($\mathbf{a}$) will be false in the resulting state and fluents that are not negated will be true. We now have all the notation and definitions we need to define a *plan*.

Definition 3.11 Given a domain theory, $\mathcal{T}^{\mathcal{I}} = \langle \mathcal{F}, \mathcal{A} \rangle$, and an initial state, s_i, we define a *plan* as a sequence of actions $p = [\mathbf{a}_1, \mathbf{a}_2, \ldots, \mathbf{a}_n]$ such that $\mathbf{a}_i \in \mathcal{A}$, and $\forall i \in [1, n-1]$, all fluents in pre($\mathbf{a}_{i+1}$) are true in the state that results from executing actions $\mathbf{a}_1, \ldots, \mathbf{a}_i$ in order starting execution in s_i.

Definition 3.12 Next, given a domain theory, $\mathcal{T}^{\mathcal{I}} = \langle \mathcal{F}, \mathcal{A} \rangle$, an initial state, $s_i \in \mathcal{S}_{\mathcal{F}}$, and a goal condition, $g \subseteq \mathcal{F}$, we define a *plan for a goal*, g as a plan, $p_g = [\mathbf{a}_1, \mathbf{a}_2, \ldots, \mathbf{a}_n]$ starting in s_i, such that all fluents in g hold after the execution of $\mathbf{a}_n$.

Definition 3.13 Further, given a plan $p = [\mathbf{a}_1, \ldots, \mathbf{a}_n]$ defined using $\mathcal{T}^{\mathcal{I}}$ and s_i, we will say that p *satisfies* an observation sequence $O = [\mathbf{o}_1, \ldots, \mathbf{o}_k]$ such that $\mathbf{o}_i \in \mathcal{A}$, $k \leq n$ and there is a monotonically increasing mapping of the elements in O to the elements in p. We denote this as $p \models_{\mathcal{I}} O$.

Note, unlike explicit representations, these definitions do not require p to be a minimum length or otherwise optimal plan for the goal. p could contain exogenous or even redundant actions. The plans people execute in the real world are often not optimal in these ways and we want to make sure that our definitions address this. Furthermore, the mapping of O to p is not required to be complete or one to one. Since we will most often be doing prefix recognition, k will often be significantly less than n preventing any complete one to one mapping. Moreover, many domains are only partially observable and therefore observations may be missing even from a complete execution of a plan prefix.

Definition 3.14 Finally, given a domain theory, $\mathcal{T}^{\mathcal{I}} = \langle \mathcal{F}, \mathcal{A} \rangle$, we define an *instance*, i, of an implicitly defined recognition problem as a tuple, $i = \langle \mathcal{T}^{\mathcal{I}}, s_i, \mathcal{G}, O \rangle$, such that:

- $s_i \in \mathcal{S}_{\mathcal{F}}$ is the initial state of the world;

- $\mathcal{G} = \{ g_1, g_2, \ldots g_n \}$ defines the possible goal conditions the agent could be trying to reach in the world. This means each g_i is a condition over the domain theories fluents, $\forall\, g_i \subseteq \mathcal{F}$. We will call states in which one of the g_i hold *goal states*; and

- $O = [\mathbf{o}_1, \ldots, \mathbf{o}_k]$ such that $\mathbf{o}_i \in \mathcal{A}$ is an ordered sequence of observed actions the agent executed.

The Plan Recognition Task

So, given an $i = \langle \mathcal{T}^{\mathcal{I}}, s_i, \mathcal{G}, O \rangle$, the plan recognizer's task is to find a plan, p_g for achieving (not necessarily optimally) a goal $g \in \mathcal{G}$, that satisfies O. This is notated by $p_g \models_{\mathcal{I}} O$.

Given this restatement of the plan recognition task, the *Goal* recognition task in implicit representation systems either identifies a single most likely goal $g \in \mathcal{G}$ that has a plan p_g that satisfies the observations, or a probability distribution over all the possible goals in $\mathcal{G}$ that reflects their conditional probability given the observation sequence. In addition, *Plan* recognizers identify a plan (or plans) that achieve the goal (or distribution) that satisfy the observations.

3.3.2   RUNNING EXAMPLE

A simple domain theory, $\mathcal{T}^{\mathcal{I}} = \langle \mathcal{F}, \mathcal{A} \rangle$, for the breakfast example:

- $\mathcal{F} = \{$ *Mug, HaveMug, CoffeeBeans, HaveCoffee, Glass, HaveGlass, Oranges, HaveJuice, Bagel, HaveBagel, CreamCheese, HaveBagelWithCheese, Bowl, HaveBowl, Cereal, HaveCereal* $\}$

- $\mathcal{A} = \{$

    GetGlass( ): $\{$ *Glass* $\} \rightarrow [\ \{$ *HaveGlass* $\}, 1.0\ ]$,

    JuiceOranges( ) : $\{$ *HaveGlass, Oranges* $\} \rightarrow [\ \{$ *HaveJuice* $\}, 1.0\ ]$,

    GetMug( ): $\{$ *Mug* $\} \rightarrow [\ \{$ *HaveMug* $\}, 1.0\ ]$,

    GrindBeans( ) : $\{$ *HaveMug, CoffeeBeans* $\} \rightarrow [\ \{$ *HaveCoffee* $\}, 1.0\ ]$,

    GetBagel( ) : $\{$ *Bagel* $\} \rightarrow [\ \{$ *HaveBagel* $\}, 1.0\ ]$,

    SpreadCreamCheese( ) : $\{$ *HaveBagel, CreamCheese* $\} \rightarrow [\ \{$ *HaveBagelWithCheese* $\}, 1.0\ ]$,

    GetBowl( ) : $\{$ *Bowl* $\} \rightarrow [\ \{$*HaveBowl*$\}, 1.0\ ]$,

    PourCereal: $\{$ *HaveBowl, Cereal* $\} \rightarrow [\ \{$ *HaveCereal* $\}, 1.0\ ]\ \}$

    DrinkCoffee: $\{$ *HaveCoffee* $\} \rightarrow [\ \{$ ¬*HaveCoffee* $\}, 1.0\ ]\ \}$

    DrinkJuice( ) : $\{$ *HaveJuice* $\} \rightarrow [\ \{$ ¬*HaveJuice* $\}, 1.0\ ]\ \}$

EatBagel: { *HaveBagelWithCheese*} → [{ ¬*HaveBagleWithCheese*}, 1.0] } }

EatCereal: { *HaveCereal*} → [{ ¬*HaveCereal*}, 1.0] } }

We use the right arrow to define actions as {pre(**a**)} → [{post(**a**), *cost*(*act*)]. For example, the above definition says that to execute the action **JuiceOranges**, it must first be the case that we have the glass and there are oranges (i.e., the fluents *HaveGlass* and *Oranges* are both true) and as a result we will have juice (i.e., the fluent *HaveJuice* is true). Reminder, icons have been provided for actions that will be used frequently in later examples.

Now consider we have the following instance, $i = \langle \mathcal{T}^{\mathcal{I}}, s_i, \mathcal{G}, O \rangle$:

- $s_i = \{Glass, Oranges, Bagel\}$ with all other fluents assigned to false.

- $\mathcal{G} = \{$ JuiceMeal, CoffeeMeal, BagelMeal, CerealMeal$\}$ such that:

   – JuiceMeal=\{ *HaveJuice*\},

   – CoffeeMeal=\{ *HaveCoffee*\},

   – BagelMeal=\{ *HaveBagelWithCheese*\},

   – CerealMeal=\{ *HaveCereal*\}.

- O = [(**GetGlass**), (**JuiceOranges**), (**GetBagel**)].

A possible output of a goal recognizer could be:

$$\langle \texttt{JuiceMeal} : 0.8, \texttt{CoffeeMeal} : 0.01, \texttt{BagelMeal} : 0.18, \texttt{CerealMeal} : 0.01 \rangle$$

indicating that JuiceMeal is the most likely goal the agent is trying to achieve. The plan that supports this conclusion is the observation sequence itself, which actually achieves the goal. CoffeeMeal and CerealMeal are not at all likely because not only have we have seen none of the actions that might contribute to it but the materials needed for it are not present. On the other hand, the goal BagelMeal is much less likely than JuiceMeal, since we have only seen one action that contributes to it, but it is still not out of the running. In the next section, we discuss a particular plan recognition system that uses an implicit representation of the domain theory.

3.3.3   PLAN RECOGNITION AS PLANNING

Citation: M. Ramírez and H. Geffner, Plan recognition as planning, *21st International Joint Conference on Artificial Intelligence*, 2009.[2]

Application Domain: International Planning Competition Domains.

Technical Approach: Using planners to map observations to goals.

[2]Ramírez and Geffner [2009]

Fundamental Contribution: Solving goal recognition using planning techniques.
Limitations: Can only reason about one plan at a time.

This work started an entire line of research called *plan recognition as planning (PRaP)* and was in fact the first plan recognition system to use an implicit representation of the domain theory. In it the authors proposed performing plan recognition by making multiple calls to an AI planing system and comparing the results. For example, given a sequence of observations, O, this approach finds two plans for each possible goal, g $\in \mathcal{G}$. The first, p_g, is a plan that satisfies the observation sequence, $p_g \models_{\mathcal{I}} O$, while achieving g. The second, $p'_g \not\models_{\mathcal{I}} O$, diverts from the observation sequence by at least one action before reaching the goal and hence does not satisfy the observation sequence. Given such a pair of plans we can compute the *marginal cost* of the observations assuming that the observations are contained within at least one optimal plan for one of the goals. The marginal cost of the observations toward the goal can be approximated as the cost difference between the these two plans: $\delta(g|O) = C(p_g) - C(p'_g)$, where the cost of each plan is computed by summing the costs of the individual actions within the plan. Remember that these costs are provided as part of the definition of each action.

Marginal cost can be seen as an indication of how consistent a given set of observations are with the optimal plan. Thus, lower marginal cost plans should be thought of as more likely explanations for the observations. To clearly see how this works we walk through the following example.

Example 3.15   Given the implicitly defined domain theory above, assume the following instance, $i = \langle \mathcal{T}^{\mathcal{I}}, s_i, \mathcal{G}, O \rangle$:

- $s_i = \{Glass, Mug, Oranges, CoffeeBeans, Bagel\}$ with all other fluents assigned to false.

- $\mathcal{G} = \{\texttt{JuiceMeal}, \texttt{CoffeeMeal}\}$ such that:

    - $\texttt{JuiceMeal} = \{HaveJuice\}$,
    - $\texttt{CoffeeMeal} = \{HaveCoffee\}$

- $O = [\,\overline{\underline{\cup}}\,(\textbf{GetGlass}),\ \textcircled{\raisebox{0pt}{}}\,(\textbf{DrinkJuice})\,]$.

To recognize which plan is being executed [Ramírez and Geffner, 2009], compute two plans for each goal for a total of four plans:

$$p_{\texttt{JuiceMeal}} \models_{\mathcal{I}} O, \ p'_{\texttt{JuiceMeal}} \not\models_{\mathcal{I}} O, \text{ and}$$
$$p_{\texttt{CoffeeMeal}} \models_{\mathcal{I}} O, \ p'_{\texttt{CoffeeMeal}} \not\models_{\mathcal{I}} O.$$

Because the action $\textcircled{\raisebox{0pt}{}}$(**DrinkJuice**) deletes the fluent *HaveJuice* from the current state of the world, the only way we can reach goal $\texttt{JuiceMeal}$ from this initial state consistent with the observations is by a plan that contains two observed $\overline{\smile}$(**JuiceOranges**) actions, one of which produces the juice that is then consumed and one that produces the juice that our goal requires.

$p_{\texttt{JuiceMeal}} = [\ \Box,\ \Box,\ \Box,\ \Box\ ].$

$p'_{\texttt{JuiceMeal}} = [\ \Box,\ \Box\ ].$

$p_{\texttt{CoffeeMeal}} = [\ \Box,\ \Box,\ \Box,\ \Box,\ \Box\ ].$

$p'_{\texttt{CoffeeMeal}} = [\ \Box,\ \Box\ ].$

The shortest plan for $p'_{\texttt{JuiceMeal}}$ simply ignores the drinking observation in the plan and adds this to its cost. Thus, assuming a unit cost of all actions, we get:

$$\delta(\texttt{JuiceMeal} \mid O) = C([\ \Box,\ \Box,\ \Box,\ \Box\ ]) - C([\ \Box,\ \Box\ ]) = 2.$$

Further:

$$\delta(\texttt{CoffeeMeal} \mid O) = C([\ \Box,\ \Box,\ \Box,\ \Box,\ \Box\ ]) - C([\ \Box,\ \Box\ ]) = 3$$

because the cost of a plan for `CoffeeMeal` that makes use of the observations is so large, `JuiceMeal` has the smaller cost. Therefore, the system would conclude both that $p_{\texttt{JuiceMeal}}$ is the goal and that the agent's plan that is shown above. Note the initial work only returned the goal but was later extended to include the plan making it a true plan recognition system. This work was also extended to handle cases where agents are not assumed to behave optimally and therefore there may be no goal with an optimal plan that satisfies the observations [Ramırez and Geffner, 2010]. This extended formulation gives a softer approximation for costs, but will still rank `JuiceMeal` to be more likely than `CoffeeMeal` in this case.

We note that like all of the algorithms that use implicit representation that we will discuss here, this work assumed that the actor is pursuing a single plan to achieve a single goal at a time. That is, such systems are unable to recognize a situation where the agent is concurrently carrying out two unrelated plans to achieve two different goals at the same time (e.g., cleaning the kitchen while doing laundry).

3.4 FINAL THOUGHTS ON IMPLICIT VS. EXPLICIT REPRESENTATION

Having formalized and given examples of implicit and explicit domain representations it is worthwhile to ask at least a few challenging questions about the differences between them and their relative merits. Intuitively, systems using implicit representations seem to require less knowledge in their domain theories, and this was an early claim of its supporters. In addition to the domain's actions, explicit domain theories must encode the plans to be recognized. However, there are subtleties to this issue that are worth discussing.

As we have already mentioned, systems that use an explicit representation often work only within an action space. This means that, while they have to identify the actions and their acceptable sequences, they may need less knowledge about the actions. For example, as we pointed out, the PHATT system does not use knowledge of the preconditions and effects rules of the actions to recognize specific sequences of actions, and it did not represent this knowledge in its domain theories. Thus, there is a possible trade off between causal knowledge of actions and sequential knowledge of plans that have to be represented in a domain theory.

Further, for real-world domains, preconditions and postconditions in action representations are often used to encode search knowledge in addition to causal knowledge of actions. That is, preconditions and even postconditions can be added to actions, not to capture causal knowledge about what will eventuate by executing the action, but rather to constrain when the action should (or shouldn't) be used in building a plan. This is often done to speed the planning process (i.e., preventing the system from searching for plans that are unacceptable). One can see preconditions with this function as just a different method of encoding the same information captured in explicit domain theories but in a manner distributed across each action. As a result, the knowledge engineering of actions in implicit domains can be quite onerous. This should further call into question our intuition that systems using explicit domain theories require more encoded knowledge.

Another common objection to explicit domain representations is to ask "where do the set of plans to be recognized come from?" While the vast majority of both kinds of systems currently use hand coded domain theories, this is a natural and legitimate question to ask where this kind of knowledge comes from and who will be encoding it. There are a number of potential answers for explicit domain theories. For example, in some domains such collections of plans already exist in the form of "doctrine" or mandated operational procedures. Consider high-risk settings where there are well known and codified sets of acceptable plans (e.g., military doctrine, large-scale manufacturing plant operating processes, flying airplanes, medical, or even legal procedures, etc.). In such settings these standard procedures often must be followed or the agent runs the risk, not only of not achieving their goals, also being censured by higher authorities (e.g., court marshal, malpractice, loosing ones operators license or other certification, etc.) if not causing tragic results. In fact, sometimes the reason these procedures have already been codified is exactly to guarantee that one can easily identify when they are being done correctly by others. Thus, recognizing when such specific procedures are not followed is an area ripe for the application of plan recognition (e.g., recognizing financial fraud or other misbehaving professionals), and there are already well agreed-upon and codified sets of acceptable plans that can be hand encoded into explicit domain theories for them.

That said, the long-term solution for where explicit domain theories come from, must rest with machine learning. While there has already been work on learning the explicit domain knowledge for some systems [Geib and Kantharaju, 2018], this is still a largely open research question and one that has many applications. In fact, there is also a large body of work on learning the domain knowledge needed for implicit domain theories as well. However, a complete treatment of this research area is well outside the scope of this book.

Finally, one the arguments in favor of implicit over explicit domain representation is try to leverage research in another area to address plan recognition. That is, to convert one research problem into another well-known research problem and make use of the second research area's results. This can be a very valuable approach to solving a problem. For example, we have already seen that explicit representations have been heavily informed by NLP. But again it is worthwhile to examine this claim at least in a little detail.

In general, the claim behind implicit domain representations is that one can reduce the problem of plan recognition to that of another "AI hard" problem namely that of planning. While some planning problems are easy to solve, it is well known that AI planning in the worst cases is computationally intractable.[3] Thus, while using AI planning algorithms may reduce one problem to another that is highly studied, it is not clear if the cost of this solution to the problem is worth paying. On the other hand, there is a large body of research on AI planning and many extensions that might be brought to bear on the plan recognition problem with this transformation.

Thus, in the end, it is not clear that one approach is better than another or even that difficult questions about them have simple clear answers. What does seem clear is that there are valuable insights to be gained from work that makes use of either representation, and that future research in both areas may reveal important insights into more efficient and expressive plan recognition systems.

Finally, it is also worth noting that in both implicit and explicit representations, many of the models used here were first formalized for AI planning research. As such, they can and often have been extended to be first-order, support continuous actions, and even stochastic actions. And some of the prior work has done this without reporting such extensions as part of their papers. Here we have reported on the papers as written, but we encourage the interested reader to contact the authors of these works to see if these systems have been extended.

---

[3]In contrast, the parsing algorithms that have informed explicit representations and plan recognition algorithms are known to have only polynomial (in this case cubic) runtimes.

CHAPTER 4

Improving a Recognizer

The previous chapter introduced implicit and explicit domain representations and how they are commonly used for goal and plan recognition. For each representation, we chose a leading concrete representation and discussed a specific piece of work that uses it. However, regardless of the type of representation is used, there are ways to improve the performance of a recognition: making it faster, improving its ability to capture complex plans, extending it to handle new types of observations, enriching the representation of its results, increasing its accuracy, and more.

4.1 MAIN IMPROVEMENT CATEGORIES

Since this book cannot discuss all the possible ways to improve a recognition process, we will focus on two general categories: speed and representational richness. Thus, a recognizer can be considered an improvement over existing works if it is able to either recognize the same plans faster, or if it is able to perform recognition in more complex settings. Unfortunately, these two qualities often conflict—increasing recognition speed often means that something is lost in terms of richness, and more elaborate recognition requires more computation time. As of the writing of this book, there is no agreed upon standard way to evaluate the trade-off between these properties. Instead of an empirical evaluation, this chapter will discuss how some systems attempt to improve recognition. Then in the summary of this chapter, we provide a table that summarizes traits of different algorithms in terms of speed, completeness, and what outputs they provide.

Chapter 3 already presented one of the most well known of these trade-offs. Algorithms using explicit representations can be faster than algorithms using implicit representations, but at the cost of only searching for plans that are within the explicitly enumerated set. Having an implicit model of plans usually requires multiple calls to an AI planner, which for some domains can be intractable while traditional NLPs can be as fast as cubic time, and for grammars with a finite yield faster still. That said the plans that systems with implicit representations finds can fall outside the enumerated plans of the explicit systems. As we have already discussed, this comparison does not always do justice to either of the families: having structure such as an explicit plan library can often prevent finding plans that are unacceptable for reasons other than strictly causal possibility. However, as a general rule-of-thumb, implicit representations may be able to find a wider array of possible plans, but can also be more costly to use than explicit representations.

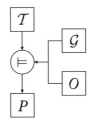

Figure 4.1: **Abstract recognition system.**

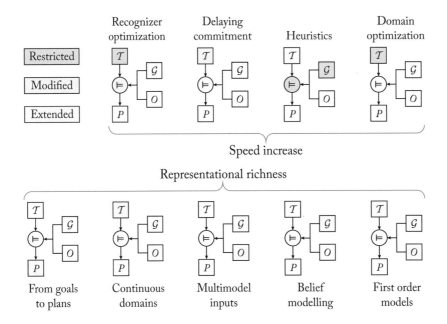

Figure 4.2: **A summary of all of the improvements presented in this chapter.**

Before we continue to describe specific systems, we remind the reader of the high-level inputs and outputs of all recognition problems. All recognizers receive as input a domain representation $\mathcal{T}$, a set of possible goals the actor might pursue $\mathcal{G}$ (either explicitly or implicitly defined), and a sequence of observations, O. The output of a *goal* recognizer is either a distribution over $\mathcal{G}$ or a single goal $g \in \mathcal{G}$. The output of a *plan* recognizer is a (possibly unary) set of plans, P, such that for each plan for a goal, g, $p_g \in P$, satisfies the observations, $p_g \vDash O$. Figure 4.1 captures this relationship as an abstract schematic of such a recognition system.

Using this model, the rest of this chapter is organized as follows: it starts with a discussion of speed and representational richness in the context of both the implicit and explicit representations presented in Chapter 3. Section 4.2 discusses speed-related improvements and efforts to make plan and goal recognition faster.

Works that handle richer representations are presented in Section 4.3. Section 4.4 discusses goal and plan recognition algorithms that provide different approaches and could not be considered as an improvement to either category. For each discussed area of improvements, we first provide a high-level introduction and references to relevant works, and then provide more detailed discussions about a selection of these works.

Figure 4.2 summarizes the improvement approaches discussed in this chapter. For each improvement, a colored component indicates that this component is changed: Red means that additional restrictions or limiting assumptions are made with respect to the component, Yellow means that the component was modified, and Green means that limiting assumptions were relaxed, or information was added. Note, changing one component may requires more changes, but we only highlight the principle component changes.

This chapter ends with two tables that summarize and provide some of the key properties of the discussed algorithms. Specifically, Table 4.1 summarizes the fine-grained properties of the algorithms while Table 4.2 describes the given algorithms in terms of completeness, soundness, and output type.

4.2   SPEED

In most cases, recognition is a component of a larger system, where the observer is required to act given the recognized goals or plans. Thus, rapid recognition is a highly desired property: a service robot providing assistance, a cyber security system that needs to detect and counter attacks, a medical system that provides critique to a physician that is following medical guidelines during an emergency, and more. There are several ways to improve the speed of a recognizer, and we now provide details about some of the key approaches. For each we will mention at least one notable work that achieved speedup using them.

4.2.1   OPTIMIZING PARTS OF THE RECOGNITION PROCESS

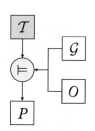

The most common approach to speed up any algorithm is to simplify or optimize a computationally expensive part of it. This works best when it is a core data structure that needs to be copied or a part of the algorithm that is frequently called. For recognition algorithms, this can be accomplished in at least three ways: restricting the applicable domains, pre-computing components offline, and simplifying supporting data-structures.

Restricting the environment can lead to significantly simpler domain representations. The challenge in doing this is to not prevent the recognition of the desired plans. For example, the work by Masters and Sardina [2017] requires significantly fewer computations than other PRaP systems, but is restricted to path planning domains.

Speed up can also be achieved by computing once, offline, and before recognition, those portions of the computation that don't depend on the observations. For example, E-Martín et al. [2015] use a plan cost model that does not require an online planner. Thus, computing the cost of each combination of actions and its influence on the recognition process can be done offline before goal recognition.

Sometimes, the simplification is more subtle. For example, Geib et al. [2008] showed that while recognition of the plan is important, the structure that used to build the plan may be redundant. Maraist [2017] took this idea further by augmenting the recognizer's parser with a shuffle operator, YR. This operator splits the recognition process into separate yet interleaved parts: observation processing and shuffling. We discuss these approaches in more detail below.

Cost-Based Goal Recognition for Path-Planning
Citation: P. Masters and S. Sardina, Cost-based goal recognition for path-planning, *Proceedings of the 16th Conference on Autonomous Agents and MultiAgent Systems*, pp. 750–758, 2017.[1]
Application Domain: Path finding benchmarks (Sturtevant [2012]).
Technical Approach: Plan recognition as planning.
Fundamental Contribution: Reasoning about the computational costs that can be saved in the online recognition process.
Constraints: Does not apply without extension to general planning domains.

This work speeds recognition by separating offline and online computation in path planning domains. A central realization behind its success is that calculating *some* optimal plan to reach goal g can be done once, offline, and before recognition. Based on this, this work refines the original goal likelihood formula:

$$\delta(g \mid O) = C(p_g \vDash O) - C(p'_g), \qquad (4.1)$$

where p'_g is the optimal plan for reaching g, which is independent of O. This formula takes the difference in costs between an optimal plan to reach g that also satisfies O, and an optimal plan with no additional constraints.

Example 4.1   While this work was originally presented for path-planning, we can use it in our running Example 3.15. Assuming a two-action observation sequence, [⬜ (**GetGlass**), 🧑 (**DrinkJuice**)], we need to compute four plans, $p_{\text{JuiceMeal}} \vDash O$, $p'_{\text{JuiceMeal}}$, $p_{\text{CoffeeMeal}} \vDash O$, $p'_{\text{CoffeeMeal}}$. As in Section 3.3 the only way we can reach a goal with a glass of juice, is if we execute two actions for making orange juice (🍊):

- $p_{\text{JuiceMeal}} = [\,⬜, \,🍊, \,🧑, \,🍊\,].$

- $p'_{\text{JuiceMeal}} = [\,⬜, \,🍊\,].$

- $p_{\text{CoffeeMeal}} = [\,⬜, \,🍊, \,🧑, \,☕, \,🫖\,].$

---

[1]Masters and Sardina [2017]

- $p'_{\texttt{CoffeeMeal}} = [\;☕,\;🍵\;]$.

Using Equation 4.1, we get that $\delta(\texttt{JuiceMeal} \mid O) = C(\;🥛,\;🍊,\;🧃,\;🥣\;) - C(\;🥛,\;🍊\;) = 2$, while $\delta(\texttt{CoffeeMeal} \mid O) = C(\;🥛,\;🍊,\;☕,\;🫖,\;🍵\;) - C(\;☕,\;🍵\;) = 3$. This means that the cost of $\texttt{CoffeeMeal}$ is higher than that of $\texttt{JuiceMeal}$. The benefit of this work in comparison to the original formulation is that $p'_{\texttt{JuiceMeal}}$ and $p'_{\texttt{CoffeeMeal}}$ can be calculated in advance, before any actions are observed.

A Fast Goal Recognition Technique Based on Interaction Estimates
Citation: Y. E-Martin, M. D. R-Moreno, and D. E. Smith, A fast goal recognition technique based on interaction estimates, *24th International Joint Conference on Artificial Intelligence*, 2015.[2]
Application Domain: Same planning domains as Ramírez and Geffner [2009].
Technical Approach: Plan recognition as planning.
Fundamental Contribution: Significantly improves runtime over the original PRaP approach.
Constraints: This work focuses on interactions between pairs of fluents, therefore interactions between three or more fluents can cause problems for this approach.

While the input and output of this problem are identical to those described in the original PRaP formulation, this work focuses on an alternative approach to calculate the costs of plans. Here, a metric called *Interaction Cost*, (IC) defines the difference between two fluents, f_1 and f_2:

$$IC(f_1, f_2) = cost(f_1 \wedge f_2) - (cost(f_1) + cost(f_2)). \tag{4.2}$$

When $IC < 0$, the cost of establishing both f_1 and f_2 is less than the sum of the costs of establishing the two independently. When $IC = 0$, the fluents are independent. When $IC > 0$, the fluent interfere with each other. Using this estimate, a plan graph is constructed such that the interaction between two possible fluents is in $\{1, 0, -1\}$, which intuitively means whether the fluents are synergistic, independent from one another, or conflicting. This graph can be constructed offline greatly speeding the algoerithm's plan cost computation and overall runtime.

Example 4.2   Figure 4.3 shows the plan graph for our running example. Assuming the observations (shown in square frames in the figure) are: $🥛$ at time $t = 0$ and $🍊$ at time $t = 2$. The numbers $(1, 0, -1)$ above the fluents and actions are the values for each fluent and action computed using the above domain theory. For example, if we assume that we start with the fluent *Oranges* being true (it is labeled as 1), then observing $🥛$(**GetGlass**) at the first time step will trigger *Glass* to be true as well.

---

[2]E-Martín et al. [2015]

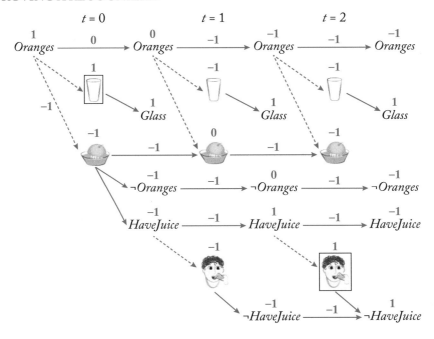

Figure 4.3: Plan graph with interaction costs for the running example.

We will show the costs required to compute $PR(\texttt{JuiceMeal})$. First, the interaction cost of $\texttt{JuiceMeal}$ without the observations is:

$$cost(\texttt{JuiceMeal}) = cost(Glass \wedge HaveJuice) \approx$$
$$cost(Glass) + cost(HaveJuice) + IC(Glass, HaveJuice) = 2 + 4 + 0 = 6. \quad (4.3)$$

Next, the interaction cost of $\texttt{JuiceMeal}$ given the observations is:

$$cost(\texttt{JuiceMeal} \mid O) = cost(Glass \wedge HaveJuice) \approx$$
$$cost(Glass) + cost(HaveJuice) + IC(Glass, HaveJuice) = 4 + 2 - 1 = 5. \quad (4.4)$$

Finally, the cost difference is: $\Delta(\texttt{JuiceMeal}, O) = cost(\texttt{JuiceMeal}|O) - cost(\texttt{JuiceMeal}) = 5 - 6 = -1.$

YAPPR: Yet Another Probabilistic Plan Recognizer
Citation: C. W. Geib, J. Maraist, and R. P. Goldman, A new probabilistic plan recognition algorithm based on string rewriting, *ICAPS*, pp. 91–98, 2008.[3]
Application Domain: Artificial plan libraries.

[3]Geib et al. [2008]

Technical Approach: Weighted model counting using string rewriting.

Fundamental Contribution: Reasoning about the components of the parse trees that are redundant to an explanation.

Constraints: Cannot reconstruct the parse trees if needed (e.g., to show how the observations are explained by the parsing process).

Yet Another Probabilistic Plan Recognizer (YAPPR) is an improvement on PHATT's probabilistic parsing approach. It uses the same pending sets and probability model based on weighted model counting of explanations generated by parsing a grammar. However, rather than using tree adjunction for parsing, it uses string rewriting. In its plan library, and its explanations, rather than using a left-most trees the grammar is rewritten back into a context free grammar that keeps just the frontiers of the trees.

This makes YAPPR grammars exactly Greibach Normal form grammars for the plans. This has the knock-on effect that output of the YAPPR system is only a collection of plan symbol strings (rather than leftward trees) that each correspond to a plan being executed. With each new observation of an agent's actions, one of the left-most (according to the partial order specified by the domain theory) non-terminal symbols in the string is replaced with a new substring of symbols given by the string grammar in the plan library. Thus, it maintains a much smaller data structure for each explanation. This significantly reduced YAPPR's runtime, however it does not address PHATT's bias for early commitment, making it still fairly computationally expensive.

Example 4.3 We use the full library from Example 3.2.2. Assume a single action was observed, ☕ (**GetMug**). Figure 4.4a shows PHATT's tree grammar for explaining the observation. Figure 4.4b shows how YAPPR rewrites this tree to a string. The root of the tree, `MakeBreakfast`, is the left-hand side of the rule. The right-hand side is just the in-order frontier of the tree including the non-terminals that are yet to be observed: `GrindCoffee` and `MakeFood`. Parsing to build up explanations is done by the derivation process described in Section 3.2.1. If an explanation contains more than one concurrent plan, YAPPR represents this as a set of strings.

YR: String Shuffling over a Gap between Parsing and Plan Recognition

Citation: J. Maraist, String shuffling over a gap between parsing and plan recognition, *Workshops at the 31st AAAI Conference on Artificial Intelligence*, 2017.[4]

Application Domain: Artificial plan libraries.

Technical Approach: Parsing enriched with a shuffling operator.

Fundamental Contribution: Allows for efficient recognition using bottom-up parsing.

Constraints: Cannot reconstruct plan trees if needed to.

---

[4]Maraist [2017]

(a)

MakeBreakfast → , GrindBeans, Makefood

(b)

Figure 4.4: YAPPR's reduction of a plan tree to a string: (a) is the original parse tree and (b) is the remaining string.

YR is a member in the PHATT, YAPPR, and DOPLAR family, in the sense that it is designed to provide probabilities for all possible goals, but does not retain detailed plan information. It extends a generalized LR parser using a shuffle operator [Earley, 1970]. The generalization of the LR allows the parser to reason about a forest of parse trees, like YAPPR which enables the consideration of multiple executed plans. The shuffle operator enables reasoning about the interleaving of the execution of the multiple partial-ordered plans. YR combines these two properties, and thus provides a competitive algorithm that is faster than YAPPR, but is not heuristic.

Example 4.4   Figure 4.5 shows that a portion of the automaton YAPPER would used to parse the breakfast example. The • symbol represents the point in the plan (sentence in the classical parsing sense) parsed so far. The running example shows two steps for drinking coffee: fetching a mug and preparing coffee beans. These two steps may occur in any order, and their steps (if they had multiple steps) would be allowed to interleave. The relationship between drinking coffee and its concurrent subtasks is modeled using a hyperedge, drawn in black.

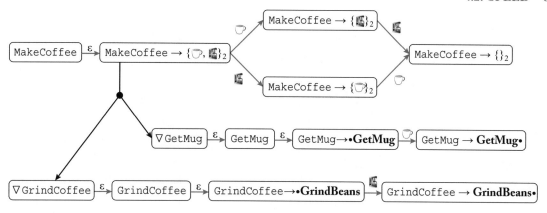

Figure 4.5: A portion of the automaton for the breakfast example.

4.2.2   DELAYING COMMITMENT

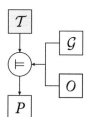

Another common approach to achieve speedup with explicit representations, is by choosing a representation that, on observing O, does not force the commitment of the recognizer to specific high level goals from $\mathcal{G}$. Delaying commitment in this way saves all of the costs of generating and maintaining a large number of hypothesis, many (if not most) of which will be inconsistent with later observations.

This delay brings up again the trade-off between speed and representational richness. The delay in commitment can significantly speed the processing of new observations faster, but at the cost of potentially delaying the recognition of the actor's goals and plans until later in the plan execution.

The first work to reason about delayed commitment uses a compact representation of the complete plan space, P, to label all potential actions that match observations from O, but verifies which of the labeled paths are consistent with the observation sequence only upon request. In Avrahami-Zilberbrand and Kaminka [2005], plans P are kept in an efficient data structure, and recognition is performed using pattern matching and consistency checks to match the observations O to plans P.

A second approach to delaying commitment was proposed by Geib [2009], by using Combinatory Categorial Grammars (CCGs). In CCGs, the grammar can be rewritten to place terminal *anchors* anywhere with in production rule. Choosing which observed action anchors which plans and sub-plans determines when the recognizer is forced to commit to specific goals.

Finally, another alternative approach to delaying commitment splits the recognition process into two parts: a bottom-up process in which observations are parsed into intermediate structures called fragments, and a top-down process that maps goals into combined fragments [Mirsky and Gal, 2016]. By postponing the second process, one can delay commitment

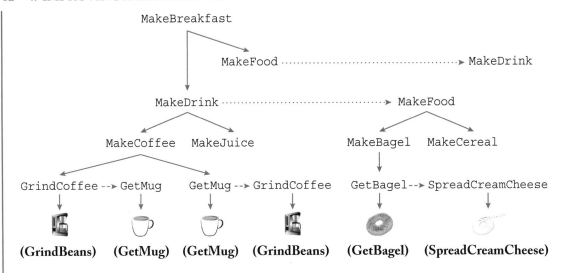

Figure 4.6: Part of SBR's plan library for the breakfast example.

to high-level goals and recognize goals locally. Below we provide additional information about each of these approaches.

SBR: Fast and Complete Symbolic Plan Recognition

Citation: D. Avrahami-Zilberbrand and G. A. Kaminka, Fast and complete symbolic plan recognition, *Proceedings of the International Joint Conference on Artificial Intelligence*, pp. 653–658, 2005.[5]

Application Domain: Artificial plan libraries.

Technical Approach: Labels over the plan library and finding consistent labels.

Fundamental Contribution: Removing the assumption that observation and actions have a bijective mapping, separating the recognition of the current state from the recognition of the actor's plan.

Constraints: Cannot reason about recursive and interleaved plans. Partial ordering requires to map all permutations.

SBR is an algorithm that actively separates the plan recognition process into three components: Matching observations to actions, matching actions to plan steps, and producing explanations. The first component, called Feature Decision Tree (FDT) is a general framework that picks up observations from the environment and decided which action (or actions) it can suit. This first step does not have a strict parallel in the other algorithms presented here, but is most similar to activity recognition. The second component matches the action to possible nodes in the plan

[5]Avrahami-Zilberbrand and Kaminka [2005]

library description, labels the nodes accordingly, and can return these nodes using a Current State Query (CSQ). Upon request, the third component can construct consistent explanations comprising of the actions marked so far, hence answering a History State Query (HSQ).

Example 4.5 SBR builds a full expansion of the plan library in advance of processing. Part of the SBR plan library for the breakfast example is shown in Figure 4.6. Solid lines represent how each NT is decomposed into other NTs and finally to actions. Dashed lines show the order by which the NTs should appear. Due to these representation, SBR needs to have a different instance for each order of a rule constituents. For example, if we wish to allow **GrindBeans** and **GetMug** to appear in any order, we'll have to represent four nodes below make coffee instead of two.

ELEXIR: Engine for LEXicalized Intent Recognition

Citation: C. W. Geib, Delaying commitment in plan recognition using combinatory categorial grammars, *21st International Joint Conference on Artificial Intelligence*, 2009.[6]

Application Domain: Artificial plan libraries.

Technical Approach: Plan recognition as parsing over combinatory categorial grammars (CCGs).

Fundamental Contribution: The use of CCGs to avoid the early commitment of the top-down parsers used before that.

Constraints: CCGs grammar construction can be knowledge intensive.

ELEXIR was explicitly built to address the early commitment problem of PHATT and YAPPR and does not require it, but can support it when desirable. Like PHATT and YAPPR, its probability model is based on weighted model counting of explanations that are produced by parsing a plan grammar. However, all of the domain specific knowledge usually in production rules is pushed into data structures related to each observable action. ELEXIR uses Combinatory Categorial Grammars (CCGs) [Steedman, 2000] to represent plans based on Combinatory Logic [Curry, 1977]. CCGs use two operators: forward slash "/" and backward slash, "\" to define functional grammatical categories to capture plans to be recognized. Each observable action can take on one of a set of complex functional categories and is said to *anchor* the plans and sub-plans captured in its respective categories. Note that each such functional category can have arguments that are *to its left* or *to its right* in the observation sequence. Parsing is accomplished by using combinators that model directional function application and composition to combine two categories into a single resulting category.

Choosing which actions anchor which plans (and subsequently begin their recognition), determines when the recognizer is forced to commit to specific goals. This flexibility in the grammar design gives the author much more control over the structures generated by each ob-

---

[6]Geib [2009]

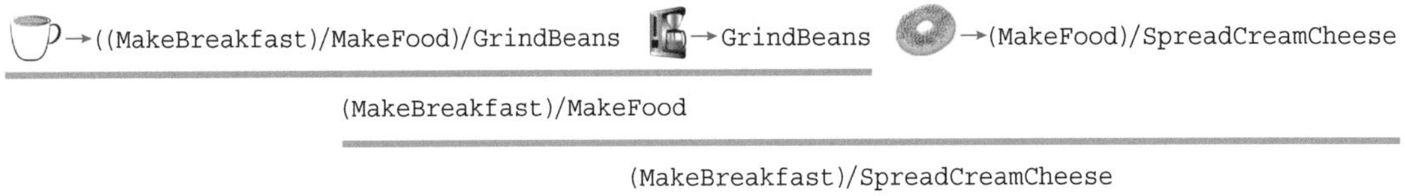

→((MakeBreakfast)/MakeFood)/GrindBeans   →GrindBeans   →(MakeFood)/SpreadCreamCheese

(MakeBreakfast)/MakeFood

(MakeBreakfast)/SpreadCreamCheese

Figure 4.7: A derivation using CCGs describing the breakfast example given the observations. It combines two categories using one rightward application (upper line) and the resulting category with a third using one rightward composition (lower line).

servation. Anchors that are late in the plan can be parsed exceptionally efficiently if the domain theory categories delay building hypothesis as long as possible. While choices of early anchors can again force the system to consider high level goals early, potentially allowing much earlier recognition but slowing the system (like PHATT and YAPPR).

Example 4.6   In order to perform plan recognition using ELEXIR, we must define a CCG based domain theory. This entails assigning sets of CCG categories to each possible action in the domain theory to capture the plan grammar. For example, in order to represent the fact that MakeBreakfast and MakeDrink and GetMug and GrindCoffee are unordered in our original example plan library, the observable action (**GrindBeans**) will have a number of categories to capture all of the possible cases for the orderings:

(**GrindBeans**)→   ((MakeBreakfast)/MakeFood)\GetMug |
((MakeBreakfast)/MakeFood) / GetMug |
(MakeDrink)\GetMug |
(MakeDrink)/GetMug |
GrindCoffee.

The first two are for the case where it is the first observed action for a MakeBreakfast plan, and the remainder are for cases where it is later in the plan. A category similar to the first one is also shown for in Figure 4.7. This is precisely because the MakeFood and MakeDrink are unordered in the original grammar. Figure 4.7 shows the initial categories assigned to the observed actions and a possible derivations for the sequence [, ,]. Note that in this case the observation of does not require the system to hypothesize another instance of MakeBreakfastwhich is a possible explanation for it. Instead, it stops at the more relevant MakeFood allowing it to play a role in the existing plan.

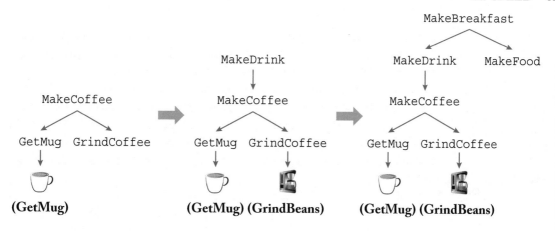

Figure 4.8: Explanation construction progress in SLIM.

SLIM: Semi-Lazy Inference Mechanism
Citation: R. Mirsky and Y. Gal, SLIM: Semi-lazy inference mechanism for plan recognition, *Proceedings of the International Joint Conference of Artificial Intelligence (IJCAI)*, 2016.[7]
Application Domain: Artificial plan libraries.
Technical Approach: Plan recognition as parsing over plan tree grammars.
Fundamental Contribution: First to combine bottom-up and top-down parsing for plan recognition.
Constraints: Can return inconsistent explanations if using just the bottom-up parsing component.

SLIM combines SBR's lazy inference that provides a full explanation only upon request (using the HSQ component) with PHATT's top-down abilities. It uses a similar incremental approach of combining observations into existing explanations like PHATT, but instead of constructing a path from the observations to the root, it constructs the path up until the first complex action that is not complete. The upper part of the parse tree is constructed upon request, like in SBR.

Example 4.7 We use the full library from Example 3.2.2. Assume a single action was observed, ☕ (**GetMug**). As seen on the left-hand side of Figure 4.8, SLIM will construct a parse tree with MakeCoffee as the root of the tree, as this is the most complex action that was not yet completed: ☕ (**GrindBeans**) still needs to be observed. Once that second observation is given, not only MakeCoffee is completed, but also MakeDrink, and thus the parse tree is built bottom-up until the MakeBreakfast action, as seen on the right-hand side of Figure 4.8.

---

[7]Mirsky and Gal [2016]

4.2.3   HEURISTICS

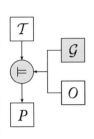

Most plan recognition algorithms are *complete* by design, meaning that if the observations, O, are consistent with some possible plan, $p \in P$ given the domain representation, the recognizer will find that plan. Another way to speed up the recognition process is to give up completeness and use heuristic functions to guide the search. To speed recognition, heuristics should be simpler to compute than the recognition process and should evaluate how close, in terms of plan execution, the actor is to each of the goals or plans. Note, using heuristics can reduce the size of the searched space, but in so doing it can discard actual solutions.

The DOPLAR [Kabanza et al., 2013] system uses a probability-based heuristic to guid its reasoning. Like the A^* heuristic search algorithm, instead of performing a Breadth-first or Depth-first search over all of the possible plans, DOPLAR estimates the likelihood of each plan to be consistent with new observations and only expands plans with high likelihood. Another heuristic approach is to estimate the likelihood of plans is using sampling. Kantharaju et al. [2019] propose the use of Monte-Carlo Tree Search (MCTS) to produce just this kind of sampling algorithm.

One of the most useful heuristic concepts in AI planning is the monitoring of *landmarks*. Landmarks are facts that must be true at some point in every valid solution plan, so they provide a concise and simple way to monitor how much the actor's execution contributes toward progressing to each goal. Pereira et al. [2017] were the first to present a heuristic goal recognizer that tracks accomplished landmarks.

When the actor is human, there are some biases that can be leveraged to heuristically prefer a subset of plans of a specific structure over others [Mirsky et al., 2017a]. For example, a person that wishes to achieve two separate goals is likely to perform as many actions as possible to achieve one goal before moving to the other, rather than jumping back and forth between the plan executions. For each of the works mentioned above, we now provide additional discussion.

Doplar: Decision-Oriented PLAn Recognizer
Citation: F. Kabanza, J. Filion, A. R. Benaskeur, and H. Irandoust, Controlling the hypothesis space in probabilistic plan recognition, *23rd International Joint Conference on Artificial Intelligence*, 2013.[8]
Application Domain: Artificial plan libraries and video games.
Technical Approach: Heuristic pruning based on weighted model counting.
Fundamental Contribution: First to use heuristics for plan recognition as parsing algorithm.
Constraints: Heuristic, hence incomplete.

---

[8]Kabanza et al. [2013]

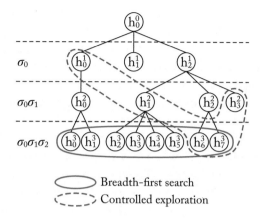

Figure 4.9: Breadth-first search compared to YAPPR's controlled exploration.

DOPLAR is a heuristic weighted model counting algorithm that limits the number of generated plan execution models in order to recognize goals quickly by computing their lower and upper bound likelihoods. The explanations are represented in a compact form similar to that of YAPPR, but its heuristic does not compute all possible explanations, but expands only the most likely ones.

Example 4.8   Figure 4.9 shows how different explanations can be developed as more observations are processed. PHATT, YAPPR, and other breadth-first methods are required to expend the whole tree, while DOPLAR chooses the next best node to expand by calculating the lower and upper bounds of goal hypothesis probabilities for each explanation.

Scaling up CCG-Based Plan Recognition via Monte-Carlo Tree Search
Citation: P. Kantharaju, O. Santiago, and C. W. Geib, Scaling up CCG-based plan recognition via Monte Carlo tree search, *IEEE Conference on Games (CoG)*, pp. 1–8, IEEE, 2019.[9]
Application Domain: Video Games.
Technical Approach: Using Monte Carlo Tree Search to sample the space of possible explanations.
Fundamental Contribution: Significant speedup in performance compared to complete parsing methods.
Constraints: As with many anytime algorithms, if not run to completion solutions can be missed.

---

[9]Kantharaju et al. [2019]

This work introduces an anytime approximation algorithm to sample the space of possible explanations for observations. While this work builds on the ELEXIR system, this technique can be used with any probabilistic. Rather than compute the entire space of possible explanations for a set of observations, it uses Monte Carlo Tree Search (MCTS) to conduct this search. As more time is given to the algorithm, its search more closely approximates the complete search.

As ELEXIR follows a plan recognition as parsing approach, searching for plans is done by parsing the observed actions. Given that most real-world plan recognition domains have significant ambiguity, this parsing process can be a very expensive search resulting from its large branching factor. This work, like other MCTS-based algorithms, uses randomized (uninformed) "rollouts" to efficiently sample the space and incrementally build a heuristic search tree to guide future rollouts and reduce the overall cost of the search. Because this is a sampling algorithm, it can be run for as much or as little time is available gathering exploring more and more of the space. If enough time is available the algorithm's informed search tree will eventually grow to encompass the entire space of valid plans, producing the same results as ELEXIR's original algorithm. If less time is available, a probabilistically guided subset of all viable plans is used to compute probabilities of ELEXIR's plans and goals.

Example 4.9   As we saw in Example 4.6, partial ordering can introduce a large number of categories into an ELEXIR style CCG grammar. While the 🖼 (**GrindBeans**) action had five possible categories in an ELEXIR CCG, the ☕ (**GetMug**) action also has the same number for the same reasons.

$$
☕(\textbf{GetMug}) \rightarrow \quad
\begin{aligned}
&\texttt{((MakeBreakfast)/MakeFood)\textbackslash GrindCoffee} \ | \\
&\texttt{((MakeBreakfast)/MakeFood) / GrindCoffee} \ | \\
&\texttt{(MakeDrink)\textbackslash GrindCoffee} \ | \\
&\texttt{(MakeDrink)/GrindCoffee} \ | \\
&\texttt{GetMug.}
\end{aligned}
$$

This means that even for the two step observations sequence [☕ (**GetMug**), intextfigGrind (**GrindBeans**)] there are twenty-five (25) possible pairings that have to be considered. This figure grows exponentially as more observations are added. MCTS tree search explores this space using the success or failure of each random rollout to guide future sampling of the space and the collection of valid observation parses. This process quickly reveals that while other parses are possible the one where both actions contribute to a single goal is the most likely.

Landmark-Based Heuristics for Goal Recognition
Citation: R. Fraga Pereira, N. Oren, and F. Meneguzzi, Landmark-based heuristics for goal recognition, *31st AAAI Conference on Artificial Intelligence*, 2017.[10]
Application Domain: Planning domains.

[10]Pereira et al. [2017]

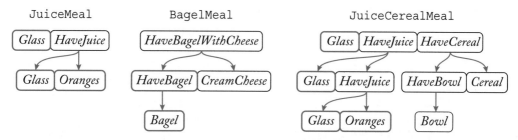

Figure 4.10: Landmarks for the goals of the running example.

Technical Approach: Plan recognition as planning.
Fundamental Contribution: This is the first work that extended the PRaP approach using heuristics.
Constraints: Heuristic, hence incomplete.

This line of work leverages AI planning, landmarks, in order to provide heuristics for plan recognition as planning. Landmarks are fluents or actions that must occur if a plan is to achieve a goal from some initial state. This work proposes two main heuristics for goal recognition. The first heuristic is to estimate proximity to each goal (the ratio between achieved and not-achieved landmarks). The second heuristic is to add weights to landmarks according to their uniqueness, and then to prefer goals with achieved unique landmarks.

Example 4.10   We use our running example to show how to compute this work's proximity estimation for goal achievement. Assume that our potential goal set is $\mathcal{G} = \{\texttt{JuiceMeal}, \texttt{BagelMeal}, \texttt{JuiceAndCerealMeal}\}$, and our initial state is $s_i = \{Glass, Oranges, Bagel\}$. Figure 4.10 depicts all the sets of landmarks per goal from our domain theory. Fluents that have been achieved have a thick outline. In the presented case, we have seen no observation yet, but many fluents are given in the initial state. The value we are interested in is $h_{prl}(g)$ which estimates the completion of a goal, g by calculating the ratio between the sum of the percentage of completion for every atomic fluent, f, that makes up g, $f \in g$. This value is:

$$h_{prl}(g) = \frac{\sum\limits_{f\in g} \frac{|\mathcal{AL}_f|}{|\mathcal{L}_f|}}{|g|}, \tag{4.5}$$

where $\mathcal{AL}_f$ is the number of achieved landmarks from observations of every f in the candidate goal g, $\mathcal{L}_f$ represents the number of necessary landmarks to achieve every $f \in g$ and $|g|$ is the number of fluents in g. For example, if we look at our initial state without any additional observations:

- $h_{prl}(\texttt{JuiceMeal}) = (1 + 2/3)/2 = 0.83$

- $h_{prl}(\texttt{BagelMeal}) = (0 + 2/4)/1 = 0.5$

- $h_{prl}(\texttt{JuiceAndCerealMeal}) = (1 + 3/5 + 0/4)/3 = 0.53$

This clearly shows that the availability of the resources in the initial state make `JuiceMeal` more likely.

4.2.4   DOMAIN OPTIMIZATION

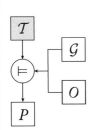

Domain optimization is any offline process that modifies the recognition domain theory such that the online recognition process is facilitated. Such offline design is usually independent of a specific sequence of observations and is typically required to guarantee that all of the original set of goals are still achievable in the new environment. Keren et al. [2014] were the first to introduce the problem of *goal recognition design* (GRD), where the aim is to find a design of the environment that minimizes the maximal progress an agent can make in a system without its goal being revealed. Wayllace et al. [2016] extended GRD to stochastic domains. Mirsky et al. [2019a] extended this work for goal and plan recognition in plan libraries. More details about these lines of work can be found below.[11]

Goal Recognition Design
Citation: S. Keren, A. Gal, and E. Karpas, Goal recognition design, *24th International Conference on Automated Planning and Scheduling*, 2014.[12]
Application Domain: Planning domains.
Technical Approach: Plan Recognition as Planning.
Fundamental Contribution: Using domain optimization in order to facilitate goal recognition.
Constraints: Assumes the observed agents only pursue one goal at a time. The original formulation assumed environments are deterministic and fully observable, but later works relaxed these assumptions [Keren et al., 2019, 2020, Wayllace et al., 2020].

Goal recognition design improves recognition by modifying the environment in order to reduce the ambiguity of the possible plans the actor can follow. It defines the *worst case distinctiveness (wcd)* of a domain theory as the maximum cost of an action sequence that is a prefix of plans to more than one goal. It then uses domain design to modify the domain theory such that *wcd* is minimized. In the original work, the actor is assumed to act optimally in a fully observable and deterministic environment. Design involves removing (disallowing) actions in the environment,

---

[11]See Keren et al. [2020] for a recent survey on goal recognition design.
[12]Keren et al. [2014]

under the constraint that the minimal cost to all goals is preserved. The original assumptions were relaxed and the original design options were extended in successive works.

Example 4.11 In the implicit domain theory from our running example, assume that our potential goal set is G = {JuiceMeal, CoffeeMeal, CoffeeAndBagelMeal}, and our initial state is s_i = {*Glass, Oranges, Mug, CoffeeBeans, Bagel*}. The *wcd* of our domain theory is two, since the sequence ☕ (**GetMug**), ⊞r (**GrindBeans**) is a shared prefix between CoffeeMeal and CoffeeAndBagelMeal. In the worst case, the goal can be recognized only after observing three actions depending on whether 🔘 (**GetBagel**) was performed. In this example, *wcd* can be minimized by imposing a partial order that guarantees that the actor can only get a bagel before doing any other action. This means that for achieving CoffeeAndBagelMeal the first action must be **GetBagel**, so the first action of the actor is guaranteed to reveal the goal.

Goal and Plan Recognition Design using Plan Libraries
Citation: R. Mirsky, Y. Gal, R. Stern, and M. Kalech, Goal and plan recognition design for plan libraries, *ACM Transactions on Intelligent Systems and Technology (TIST)*, 10(2):1–23, 2019.[13]
Application Domain: Artificial plan libraries and the Monroe disaster management corpora [Blaylock and Allen, 2005].
Technical Approach: Rule pruning for domain optimization.
Fundamental Contribution: Extending the recognition design paradigm to plan libraries.
Constraints: Can only optimize the domain assuming that one plan is executed at a time. Rule pruning in explicit representations is harder to enforce in the environment compared to the action pruning used in implicit representations.

This work extends the Goal Recognition Design problem presented by Keren et al. [2014] to plan libraries. As in the planning case, the wcd of a plan library is defined to be the length of the longest action sequence that is a prefix of optimal plans for more than one goal. It also introduces the term of *worst case plan distinctiveness (wcpd)* of a plan library, which is the length of the longest action sequence that is a prefix of any two optimal plans, even for the same goal. This work shows that in common plan libraries used in plan recognition plans for the same goal are harder to distinguish than plans for different goals.

Example 4.12 In the explicit domain theory for our running example, the *wcd* of our plan library is 0, since there is only one goal that is shared among all plans - MakeBreakfast. However, the *wcpd* of the same plan library is 3. Remember that MakeDrink can be executed before MakeFood, due to the derivation rule

{MakeBreakfast → MakeDrink, MakeFood | MakeFood, MakeDrink}.

---

[13]Mirsky et al. [2019a]

This means that in the worst case, the observer will have to first observe two basic actions that construct MakeDrink (e.g., ⏋ (**GetGlass**) and 🥣 (**JuiceOranges**)) and only when the actor sees ◉ (**GetBagel**) or ⌣ (**GetBowl**), the plan prefix is distinct from other plans. Similarly to the implicit case, removing the derivation rule that allows MakeDrink to proceed MakeFood, will cause the *wcpd* of the domain to be reduced to 0.

4.3   REPRESENTATIONAL RICHNESS

All of the works we discussed in this chapter so far have focused on improving a recognizer's speed. Next, we will discuss various works attempts to enhancing the recognizer's representational richness. Works in this area either provide a more fine-grained analysis of the observations, a more elaborate output, consider more challenging domains, or more versatile inputs than simple observations.

4.3.1   FROM GOALS TO PLANS

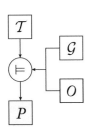

One of the first such improvements one can imagine is extending a goal recognizer to output plans. Using an implicit representation, this would mean that in addition to computing the probability of the most likely goal, also producing a plan for that goal. This is not a trivial transformation. Even if one goal is very likely to be the actor's goal, there may be several plans to reach it, which means an extended goal recognizer would need to account for these different plans, as proposed by Sohrabi et al. [2016].

Plan Recognition as Planning Revisited
Citation: S. Sohrabi, A. V. Riabov, and O. Udrea, Plan recognition as planning revisited, *Proceedings of the International Joint Conference on Artificial Intelligence*, pp. 3258–3264, 2016.[14]
Application Domain: Planning domains with missing and redundant actions.
Technical Approach: Plan recognition as planning.
Fundamental Contribution: Moving PRaP research into the realm of plans rather than goals.
Constraints: More computationally expensive than other PRaP approaches as it requires finding greater than two plans per goal.

This work has three main contributions: (1) the ability to explicitly reason about missing and noisy observations that create a bias from an optimal plan; (2) the ability to reason about observations in the form of fluents, not actions; and (3) the ability to recognize plans in addition to goals. It compares to the k-best plans (or diverse plans) for each goal g $\in \mathcal{G}$. Given O and g, the cost of plan $p_g \vDash O$ is defined as:

$$cost_{g,O}(p_g) = cost(p_g) + b_1 M_{g,O}(p_g) + b_2 N_{g,O}(p_g),\tag{4.6}$$

where $M_{g,O}(p_g)$ is the number of missing observations and $N_{g,O}$ is the number of noisy observations. This cost is calculated for k-best plans per goal, and accumulated to get the goal's likelihood.

Example 4.13   We use the domain theory used in the running Example 3.3.2. Assume two actions were observed, 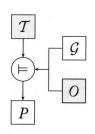 (**GetGlass**) and (**DrinkJuice**). We wish to reason about the probability of JuiceMeal being the true goal of the actor. We look at two plans that achieve JuiceMeal:

$$p_1 = [\; , \; , \; , \;].$$

$$p_2 = [\; , \;].$$

In the first plan, the action had an effect of removing *HaveJuice*, so there should be another action to achieve that fluent back. The second plan is also a valid explanation for the observations, if we assume that was a noisy observation. Using these two plans, we can calculate the following values (assume a unit cost of all actions and that the weights b_1, b_2 values are 1):

$$cost_{\text{JuiceMeal},O}(p_1) = cost(p_1) + b_1 M_{\text{JuiceMeal},O}(p_1) + b_2 N_{\text{JuiceMeal},O}(p_1) = 4 + 1 + 0 = 5.$$

$$cost_{\text{JuiceMeal},O}(p_2) = cost(p_2) + b_1 M_{\text{JuiceMeal},O}(p_2) + b_2 N_{\text{JuiceMeal},O}(p_2) = 2 + 0 + 1 = 3.$$

Using a similar calculation for k different plans, this approach computes the k-best plans to achieve each goal. In this simple example and given that $b_1 = b_2 = 1$, we get similar results to the other algorithms (for any k we choose), that JuiceMeal is the most likely goal.

4.3.2   CONTINUOUS DOMAINS

As discussed in Chapter 2, the recognition environment can often be modeled either as a set of discrete or continuous actions. When a continuous environment is modeled using discrete actions, recognizers often rely on some preliminary component to *tokenize* the behavior of the actor into a set of discrete observations. However, the quality of the plan or goal recognition can depend greatly on the quality of the tokenizer, and may even need to work in concert with it. For example, Vered and Kaminka [2017] showed that in continuous domains, for any chosen discretization, there is a plan that is ambiguous under it, and consequently there is a need to discretize the continuous environment into separate states only after the observations are given.

Heuristic Online Goal Recognition in Continuous Domains

Citation: M. Vered and G. A. Kaminka, Heuristic online goal recognition in continuous domains, *Proceedings of the International Joint Conference on Artificial Intelligence*, 2017.[15]

Application Domain: 3D navigation domain in simulations and robot navigation.

Technical Approach: Plan recognition as planning.

Fundamental Contribution: Discretization of the domain after the set of observation is given.

Constraints: The proposed heuristics can be severely affected by partial observability or noise.

The main contribution of this work is the discussion about continuous domains. It shows theoretically that for every granularity chosen, there exists a goal recognition problem such that goals in that problem are indistinguishable in the discreticized domain, yet distinguished in the continuous domain. This work proposes an approach for plan recognition, called *mirroring*, that allows the observer to discreticize the domain after the observations are given. Vered and Kaminka also propose two heuristics inspired from mirroring neurons in order to solve a continuous plan recognition problem heuristically.

Example 4.14   As this work focuses on continuous domains and our running example is discrete, we will adapt the example presented in Vered and Kaminka [2017] to our breakfast story. Imagine that Alice reaches out to take either a cup (`JuiceMeal`) or a mug (`CoffeeMeal`) from the shelf. If we choose the ϵ granularity shown in Figure 4.11, the blue arrow depicts a specific trajectory of Alice's hand that is ambiguous given ϵ. This example illustrates why for any domain discretization of size ϵ chosen offline (before observing any plan executions) there might be a trajectory that is ambiguous between two goals.

4.3.3   MULTIMODAL INPUTS

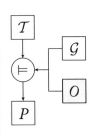
Sometimes additional communication or input channels can be used with the observations to gain knowledge about the actor's goals or plans. When using an explicit communication channel, the recognition is no longer considered *keyhole* recognition, as the observer can interact with the actor to elicit more information. While there is some potential benefit from additional input modalities, it can require the recognizer to execute additional actions beyond observing, such as asking the actor a question [Mirsky et al., 2018], or acting in a way that will affect the beliefs of the actor or the observer itself [Shvo and McIlraith, 2020b]. Alternatively, even if the recognizer remains a passive observer, observations might be in multiple modes or forms, such as gaze, in coordination with observed actions [Singh et al., 2018].

---

[15] Vered and Kaminka [2017]

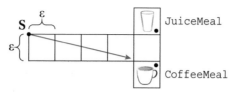

Figure 4.11: **A discretization of a goal recognition domain and an ambiguous problem instance.**

4.3.4   BELIEF MODELING

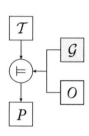

The recognition domain can also be extended by considering the mental *context*. For example, in our breakfast domain, if the robot knows that there are no oranges in the kitchen (*Oranges* is False), it should not consider plans that require (**JuiceOranges**). However, if Alice does not know that there are no oranges, she might still be executing a plan with as a future step. Only a recognizer with an explicit model of Alice's beliefs can construct a plan that is valid in that context [Fisac et al., 2020, Gmytrasiewicz and Doshi, 2005, Shvo et al., 2020].

Epistemic Plan Recognition
Citation: M. Shvo, T. Q. Klassen, S. Sohrabi, and S. A. McIlraith, Epistemic plan recognition, *Proceedings of the 19th Conference on Autonomous Agents and MultiAgent Systems (AAMAS)*, 2020.[16]
Application Domain: Epistemic planning benchmarks.
Technical Approach: Combining epistemic planning and plan recognition.
Fundamental Contribution: Bridging the gap between epistemic logic and plan recognition.
Constraints: The observer is assumed to have full observability, and its knowledge of the world is assumed to be correct.

This work builds on the idea that the recognition process is inherently epistemic (ie. it is determined by the beliefs of the observer about the beliefs, plans, and goals of the actor). The work therefore endorses the view (espoused by Pollack [1986] and others) that plan recognition may require the observer to assume the actor's unique perspective, including the latter's possibly incorrect beliefs, in order to successfully recognize the actor's plans and goals.

The work appeals to epistemic logic to explicitly represent the beliefs of the observer about the world and, importantly, about the actor's beliefs and capabilities. By modeling the observer's beliefs about the actor's beliefs as first-class elements of the plan recognition process, the work is able to recognize the actor's *epistemic* goals—related to changing the actor's own (or another

---

[16]Shvo et al. [2020]

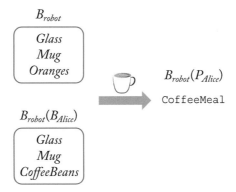

Figure 4.12: Alice's beliefs from an observer's perspective.

agent's) state of belief or knowledge. Enabled by *epistemic planning* research [Bolander and Andersen, 2011, Muise et al., 2015, Petrick and Bacchus, 2002], the work proposes a planner-independent computational realization of epistemic plan recognition as epistemic planning.

Example 4.15   Figure 4.12 shows an example of epistemic plan recognition, with the new notation B_{agent} to denote that some fluent is believed to be true by that agent. In the depicted example, the observer (the robot) correctly knows that there are no available coffee beans, while also believing that the actor (Alice) *falsely* believes that they are available. Alice is observed executing ☕ (**GetMug**) which leads the robot to reason that Alice's full plan is to have coffee. However, the robot can further reason that this plan is *ill-formed* since it will achieve the goal CoffeeMeal from Alice's perspective (due to her false belief) but will *not* achieve the goal from the robot's perspective, due to there being no available beans in the kitchen.

4.3.5   FIRST-ORDER MODELS

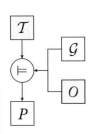

Another way to increase the richness of a plan recognizer is to model the environment using languages with more expressive power (e.g., the transition from propositions to predicates in PDDL, or using a grammar with parameters such as attribute grammars to model the environment). Using a first-order model enables the recognizer to compactly represent different instances of similar entities. For example, consider our breakfast example where the actor can pick up two objects at the same time. Using propositions to model this scenario, we would need to consider all combinations $\{Oranges + Glass, Oranges + Mug, Glass + Mug \ldots\}$ and have a unique action for each of these pairs. Using first-order logic, one the other hand, enables a more abstract representation using a single action to cover all of these cases by using variables (e.g., **PickUp**(obj_1, obj_2).) In terms of performance, a goal or plan recognizer that uses a first-order model can be more efficient, as it enumerates a smaller set of possible actions, but it might require to check the consistency of the grounded values between observations and action instances [Mirsky et al., 2017a].

CRADLE: Cumulative Recognition of Actions and Decreasing Load of Explanations

Citation: R. Mirsky, Y. Gal, and S. M. Shieber, CRADLE: An online plan recognition algorithm for exploratory domains, *ACM Transactions on Intelligent Systems and Technology (TIST)*, 8(3):1–22, 2017.[17]

Application Domain: Educational Software.

Technical Approach: Heuristic plan recognition as parsing.

Fundamental Contribution: Reasoning about practical recognition from human-generated observation sequences.

Constraints: Heuristic, can be hard to encode.

CRADLE was developed to handle real world domains, especially Exploratory Learning Environments (ELEs) where the actor can make mistakes or execute redundant actions. CRADLE is a heuristic algorithm that extends PHATT by allowing: (1) parameterized actions, (2) exogenous actions, and (3) filtering unlikely explanations. CRADLE can also be viewed as a parser for *Exploratory Grammars*, such a grammar is highly expressive and allows partial ordering, parameterized actions and interleaving of plans. This last property makes it stronger than context-free, as it can recognize languages that express: $a^n b^n c^n$.

Example 4.16   Imagine that there are several mugs in the kitchen. It is clear that if Alice picks up one mug, she is not likely to make coffee in another. This distinction is not particularly important in small settings like our kitchen example, howerver, in real world domains, enumerating all possible instances of all objects in all actions in a domain is infeasible. Therefore, CRADLE parameterizes actions. We will revise the actions in the full library from Example 3.2.2, such that select actions and non-terminals have a parameter for an identifier for the mug. For example, **GetMug**($mugID$=1) is an instance of the action for getting mugs whose ID equals one. Further, one of the derivation rules for making coffee becomes:

```
MakeCoffee(mugID) →
        GetMug(mugID), GrindCoffee [MakeCoffee : mugID = GetMug : mugID]
```

The added notation in square brackets asserts that in any derivation, the value of the MakeCoffee's $mugID$ parameter must be the same as that of GetMug's. This forces CRADLE to never produce explanations in which an observation of **GetMug**($mugID$=1) is part of a plan to make coffee in some mug other than one.

Given these changes to the plan library, assume two observed actions, **GetMug**($mugID$=1) and **GrindBeans**. First, CRADLE will update all nodes in the generated

---

[17]Mirsky et al. [2017a]

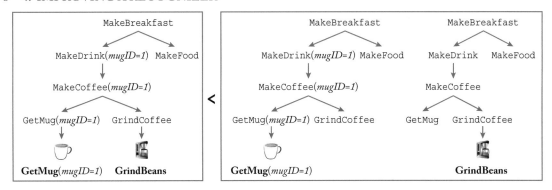

Figure 4.13: CRADLE prefers explanations that are more compact and coherent.

parse trees by binding the relevant parameters. As a result, two possible explanations for these observations can be seen in Figure 4.13. Notice that in the right-most tree, the value of the parameter $mugID$ is not set. Since, the left-hand side explanation is more compact, has fewer trees and open frontier items, CRADLE will discard the right-hand side explanation.

4.4   OTHER APPROACHES

While this chapter has covered a number of the ways in which a recognizer can be improved, it is definitely not a comprehensive list. There are still many creative, unique, and surprising ways to perform plan and goal recognition. For example, in this book we focus mostly on keyhole recognition, where the actor's plan is not affected by the presence of the observer. However, a large number of plan and goal recognizers were designed specifically for the recognition of adversaries, who might try to hide or obfuscate their goals [Geib and Goldman, 2001, Lisỳ et al., 2012].

An additional different approach for goal and plan recognition corresponds with models that are more common for activity recognition—the HMM [Russell and Norvig, 2003]. These algorithms leverage the structure of an HMM to recognize hidden goals with potentially complex plan structures given a sequence of observations [Bui, 2003].

Further, there has been excellent work based on viewing the problem as one of constraint satisfaction [Gal et al., 2012]. In this case, the observations are used as constraints on the plans that are valid. Another work along these lines is the work by Levine and Williams [2014], whose useful and rich breakfast making domain we've extended and discussed throughout this book.

Game-Theoretic Approach to Adversarial Plan Recognition

Citation: V. Lisý, R. Píbil, J. Stiborek, B. Bošanský, and M. Pěchouček, Game-theoretic approach to adversarial plan recognition, *Proceedings of the 20th European Conference on Artificial Intelligence*, pp. 546–551, 2012.[18]

Application Domain: Network attack domain.

Technical Approach: Game theory, finding an equilibrium in a game.

Fundamental Contribution: Looking at the recognition task as a goal of a player in an adversarial game.

Constraints: Can only reason about plans that are valid given the game definitions.

This work focuses on adversarial plan recognition, where the actor actively tries to avoid detection. This situation is modeled using a game-theoretic framework. This work defines the adversarial plan recognition problem as an imperfect-information extensive-form game between the observer and the actor, where (1) actions are simultaneous, (2) the actor has a set of actions to choose from, and (3) the observer has a set of classes to choose from. This work shows that with perfect knowledge, a solution to this problem is a Nash equilibrium. Outside the perfect knowledge case, this work proposes a novel algorithm that approximates the optimal solution using Monte Carlo sampling.

A General Model for Online Probabilistic Plan Recognition

Citation: H. H. Bui, A general model for online probabilistic plan recognition, *Proceedings of the International Joint Conference on Artificial Intelligence*, 3:1309–1315, 2003.[19]

Application Domain: Surveillance domain.

Technical Approach: Particle Filters over Abstract Hidden Markov Memory Model.

Fundamental Contribution: A novel model for hierarchical plan representation.

Constraints: Cannot handle interleaving of multiple plans.

While some of the other systems we have discussed here are probabilistic, this work is the first to use Hidden Markov Models (HMMs) for general plan recognition. Further this work provided a powerful and early treatment of hierarchy for HMMs in the form of Abstract Hidden Markov Memory Models (AHMEMs). AHMEMs represent hierarchical plans at varying levels of abstraction with each abstraction layer represented as an MDP.

Imagine an agent that is navigating a one thousand by one thousand celled grid world to deliver a package to one of several destinations. Given probabilistic models of the robot's locomotion system, we could imagine learning a probabilistic policy to provide the correct action for the robot to execute, depending on its goal, for every square in the grid. This would be expensive and time consuming.

---

[18]Lisỳ et al. [2012]
[19]Bui [2003]

In contrast we could imagine breaking this complete grid into a covering set of ten by ten sub-grids, learning a policy for each sub-grid, and learning a policy that treats each of the sub-grids as one location in a one hundred by one hundred grid thereby capturing a more abstract version of the original problem. We can even imagine doing this again to produce a three level abstraction with a top level abstract ten by ten grid. If the levels of the abstraction are connected such that finishing one of the sub-problems correctly triggers a state change in the higher level grid, we could view such a hierarchy of spaces as capturing the more traditional hierarchical plan structures that other systems have used. In such a system, the state of the execution of the robot's plan is described by the set of robot's states within the hierarchy, one for each level. This work describes building just such a hierarchy of HMMs including the required method of connecting the levels. Once the policies are learned for each of the abstract problems, this paper describes how to use this structure and a sequence of observations to probabilistically recognize which of the learned policies a robot agent it following using a Rao-Blackwellised Particle Filter as an approximate inference method.

Example 4.17   Figure 4.14 shows an example use for an AHMEM. On the left is a description of the environment, and two possible trajectories a person can take—one follows an optimal plan execution, and the other has a change of plans (the person goes to the mug, but then goes to the oranges and cup). On the right, there is a representation of two layers of abstractions. The upper three sets of arrows describe the person's transitions between different locations and their respective probabilities. The lower two sets of arrows describe transitions on a more abstract level, where the complex tasks are being executed.

This is very impressive work, both in the formalization of AHMEMs themselves and plan recognition in terms of them. Further, while it is able to probabilistically recognize complex policies, it has two notable limitations. First, HMMs and by extension AHMEMs use propositional representations of the world. This can mean a significant increase in the size of the representation of the domain model. Beyond a certain size of worlds this necessitates using heuristic methods to both solve the policies and recognize them. More troubling, if we are to consider multiple concurrent plans, the propositional representation requires explicitly enumerating all of the possible sets of plans (and their parameters) that could be executed at each of the levels of the hierarchy. This again results in a significant expansion of the domain models forcing increases in runtime and the need for approximation potentially reducing accuracy.

Concurrent Plan Recognition and Execution for Human-Robot Teams
Citation: S. J. Levine and B. C. Williams, Concurrent plan recognition and execution for human-robot teams, *24th International Conference on Automated Planning and Scheduling*, 2014.[20]
Application Domain: Robotic arm that assists in kitchen assignments.

---

[20]Levine and Williams [2014]

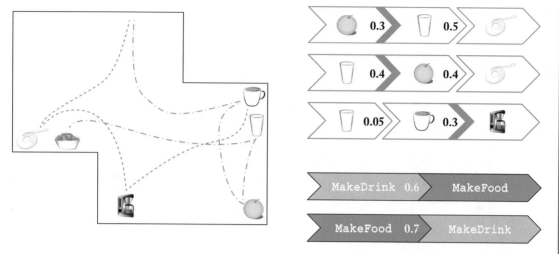

Figure 4.14: Left: an example environment with a person's trajectory; Right: a partial representation of possible transitions in two-layers of abstractions, such that each number represents the likelihood of the action to its right to succeed the action to its left.

Technical Approach: Solving a satisfiablity problem using Assumption-based Truth Maintenance System (ATMS).[21]

Fundamental Contribution: Augmenting the recognition process into the observer's planning.

Constraints: The observer is assumed to have full observability, and its knowledge of the world is assumed to be correct.

This work focuses on interleaving plan recognition with planning, by reasoning about the actor's choices and temporal constrains of the environment. The underlying representation is a contingent plan with temporal constraints, where the actions are compiled into PDDL actions, and a solution is a PDDL-based plan that is consistent with the time constraints. This system uses an Assumption-based Truth Maintenance System (ATMS) to solve this planning/plan recognition problem.

Example 4.18   Figure 4.15 again shows our breakfast example depicted using a temporal task network (TPN). In this case, circles denote events, double circles denote controllable choices (made by the observer, the robot), and shaded double circles denote uncontrollable choices (made by the actor, Alice). Each action is labeled by the length it takes to execute it (such that [x, y] means that the action takes between x to y minutes to execute). Assume that the total time Alice has to make breakfast (e.g., CoffeeAndBagelMeal or JuiceAndCerealMeal) is 7 min. If the robot observes Alice pick up a mug (◡), it knows it should help by grinding the beans

----
[21] de Kleer and Reiter [1987]

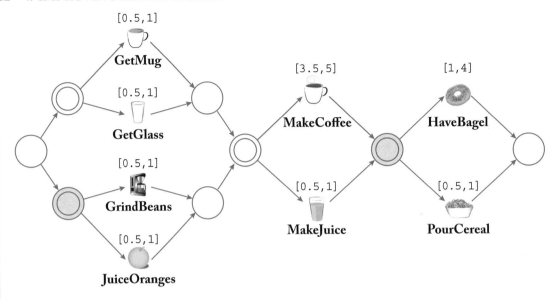

Figure 4.15: A plan network for making breakfast.

rather than bringing oranges, and even more important, it should make cereal instead of a bagel with cream cheese, since there is no time to make both a cup of coffee and a bagel in 7 min.

4.5 SUMMARIZING TABLES

In this chapter we discussed several approaches to improve a goal or plan recognition process, either by speeding or enhancing the recognition process. The main capabilities of the algorithms and techniques presented in this chapter are summarized in Tables 4.1 and 4.2. The first summarizes the assumptions made about the environment, actor, and observer. The second provides a summary of each algorithms properties.

Table 4.1: Environment, actor, and observer assumptions per algorithm. We will use a dot, "•", to indicates the algorithm makes the assumption, and a circle, "o", to indicate a more complex relationship.

Algorithm	Environment		Actor			Observer	
	Multiple goals	Plan interleaving	Keyhole	Partial observations	Noisy observations	First order domain models	Partial ordering
Gmytrasiewicz and Doshi [2005]				•	•		•
Ramirez and Geffner [2010]			•	•	•		•
E-Martín et al. [2015]			•	•			•
Sohrabi et al. [2016]			•	•	•	o	•
Masters and Sardina [2017]			•	•	o		•
Vered and Kaminka [2017]			•	•	o	o	•
Singh et al. [2018]			•	•	o		•
Pereira et al. [2019]			•	•	o	o	•
Shvo et al. [2020]			•	•	•		•
Shvo and McIlraith [2020b]			•	•	o		•
Blaylock and Allen [2005]			•				
Avrahami-Zilberbrand and Kaminka [2005]	•		•	o	•		
Geib et al. [2008]	•	•	•				•
Geib [2009]	•	•	•			o	•
Gal et al. [2012]	•	•	•	o	•		•
Kabanza et al. [2013]	•	•	•	o			•
Mirsky and Gal [2016]	•	•	•	o			•
Maraist [2017]	•	•	•	o			•
Mirsky et al. [2017a]	•	•	•	•	•		•
Mirsky et al. [2018]	•	•	•	o	o	•	•
Kantharaju et al. [2019]	•	•	•				•
Bui [2003]			•	•	o		
Lisy et al. [2012]		•	•	•			•
Levine and Williams [2014]	o		•			•	•

Table 4.2: Algorithmic properties of the presented algorithms. We use the abbreviation Dist. for *distribution over goals*, Exp. for *explanations*, CE. for *compact explanations*, and Exc. for *execution*. A dot, "●", indicates the algorithm has the property, and a circle, "○", to indicate it supports the property in some cases.

Algorithm	Output type	Complete	Sound	Speed	Predictions	Complete explanations
Gmytrasiewicz and Doshi [2005]	Dist.	●	●			
Ramírez and Geffner [2010], Ramírez and Geffner [2009]	Dist.	○				
E-Martín et al. [2015]	Dist.	○		●		●
Sohrabi et al. [2016]	Dist. + Exp.	○			●	●
Masters and Sardina [2017]	Dist.	○		○		●
Vered and Kaminka [2017]	Dist.			●		●
Singh et al. [2018]	Dist.	○		●		
Pereira et al. [2019]	Dist.			●		●
Shvo et al. [2020]	Dist.	○	○			
Shvo and McIlraith [2020b]	Dist.	○	○			
Blaylock and Allen [2005]	Dist.	●		●	●	○
Avrahami-Zilberbrand and Kaminka [2005]	Exp.	●		●		●
Geib et al. [2008]	Dist. + CE.			●	○	
Geib [2009]	Dist. + Exp.	●			●	●
Gal et al. [2012]	Exp.		●	●	●	●
Kabanza et al. [2013]	Dist. + CE.			●	○	
Mirsky and Gal [2016]	Dist. + Exp.	●		○	○	○
Maraist [2017]	Dist. + CE.	●		●	○	
Mirsky et al. [2017a]	Dist. + Exp.			●	●	●
Mirsky et al. [2018]	Dist. + Exp.	●	○	○	●	●
Kantharaju et al. [2019]	Dist. + CE.			●	○	●
Bui [2003]	Dist.	○		○	●	
Lisý et al. [2012]	Dist.	○		●		
Levine and Williams [2014]	Exp.		●	●	●	

CHAPTER 5

Future Directions

Unlike early work that treated plan recognition as just another form of inference, we now know more about what can and cannot be efficiently represented in and reasoned about within a recognition model. Regardless of which representation is used, the main challenge that still hinders building more robust and efficient goal and plan recognition tools is the complexity of the required knowledge, representations, and inference. Most current approaches rely on a significant portion of the representation to be hand-coded. This limitation highlights a set of open challenges we believe the plan recognition community must address moving forward. In the rest of this chapter, we discuss some of the challenges that we think are the next milestones for the community to conquer, and present some preliminary efforts that have been made toward them.

5.1　INTEGRATING DATA-DRIVEN TECHNIQUES

In this book, we focused on symbolic approaches to recognition problems. Recent successes in machine learning have focused on distributed representations making them difficult to directly apply to the knowledge needed for the systems discussed here. Thus, the use of data driven approaches in recognition remains limited. Instead, there are some examples of more symbolic learning that are encouraging: Kantharaju et al. [2019], Kim et al. [2018], and Maynard et al. [2019], but these are just the first important steps in a long road. Plan recognition problems are context-dependent, and in order to elicit meaningful plans from long, real-world sequences of observations, there is a need for underlying structure [Bengio]. This suggests that learning approaches for plan recognition may still require some human intelligence to complement their strength [Kaelbling, 2019].

　　Recent work has explored learning representations of the actor [Bisson et al., 2015, Geib and Kantharaju, 2018, Lioutikov et al., 2020, Pereira et al., 2019, Segura-Muros et al., 2018], While other work learns (1) to recognize the goal of an overseer and (2) how to build a plan such that a join goal is achieved.

5.2　INTEGRATING RECOGNITION AND EXECUTION

As we have already argued, recognition is usually motivated by the desire to respond to the inferred plans. Just as our example robot wanted to recognize Alice's plans to bring her tools and necessary ingredients, a network intrusion detection system needs to identify attacks and their objectives as early as possible so that limited defensive resources can be effectively allocated. Therefore, the second future challenge we highlight is the need to integrate plan and goal recognition into complex systems.

Several recent projects have worked to bridge the gap between recognition and acting in multiagent systems [Freedman and Zilberstein, 2017, Freedman et al., 2019, Levine and Williams, 2014]. A further list of other ongoing integration endeavors includes, but is not limited to:

- integrating interaction during the recognition [Alford et al., 2015, Bisson et al., 2011, Freedman and Zilberstein, 2017, Freedman et al., 2019, Mirsky et al., 2018, Shvo and McIlraith, 2020a, Sreedharan et al., 2018];

- combining activity and goal recognition [Freedman et al., 2014, Granada et al., 2020, Sukthankar and Sycara, 2005]; and

- multiagent recognition [Laviers et al., 2009, Shum et al., 2019, Sukthankar and Sycara, 2008].

Still, others have looked at plan recognition as a way for agents to explain their motivation to other agents; so-called "explainable AI." In this approach, an agent uses plan and goal recognition to explain its own actions to others: Chakraborti et al. [2017, 2019], Hadfield-Menell et al. [2016].

5.3   PLAN RECOGNITION IN THE WILD

The application of plan recognition in real-world settings has always identified new challenges. With the explosive growth of personal digital assistants, automated systems, and an ever-increasing number of embedded and "smart" controllers in our daily lives, there is a great, and ongoing opportunity for plan and goal recognition researchers to motivate and move their work into the real world. This will help us to understand the ways in which current models and methods need to be extended, as well as demonstrate the crucial relevance of our research area.

There is already much ongoing work using plan and goal recognition in:

- medical applications [Amir et al., 2013, Goldstein and Shahar, 2013, Roy et al., 2009];

- utility robots [Jiang et al., 2018, Murakami et al., 2002, Saran et al., 2018, Talamadupula et al., 2014];

- educational software [Amir and Gal, 2013, Conati et al., 1997, Gal et al., 2012];

- cyber security [Durkota et al., 2015, Geib and Goldman, 2001, Lisý et al., 2012, Mirsky et al., 2019b, Qin and Lee, 2004]; and

- user interfaces [Armentano and Amandi, 2007, Horvitz et al., 1998, Laviers et al., 2009, Liao et al., 2005, Mirsky et al., 2017b, Peng et al., 2016, Synnaeve and Bessiere, 2011], and many more.

5.4   FINAL REMARKS

Finally, the challenges and applications presented above are far from being a complete list, but hopefully they can provide a curious reader with a starting point to look for additional work, problems, and challenges relevant to their own research interests. While we are aware that nothing dates a textbook quite so much as internet links with the potential to "go stale" as the research community evolves, if we were able to inspire some of our readers to continue to work in this research area, we would strongly encourage them to use one or more of the many online resources of this community. Currently, of particular note are the following:

- PLANREC: The Plan Recognition Resources website at:
  http://www.planrec.org/.

- Our Plan and Goal Recognition tutorial that prompted this book at:
  http://www.planrec.org/Tutorial/Resources.html.

- PAIR: The yearly workshop of this wonderful community:
  http://www.planrec.org/PAIR/Resources.html.

- The PAIR mailing list at:
  https://groups.google.com/g/plan-rec.

Bibliography

Stefano Albrecht and Peter Stone. Autonomous agents modelling other agents: A comprehensive survey and open problems. *Artificial Intelligence*, 258:66–95, 2018. DOI: 10.1016/j.artint.2018.01.002 8, 24

Ron Alford, Hayley Borck, Justin Karneeb, and David W. Aha. Active behavior recognition in beyond visual range air combat. *Technical Report*, Naval Research Lab Washington, DC, 2015. 86

James F. Allen and C. Raymond Perrault. Analyzing intention in utterances. *Artificial Intelligence*, 15(3):143–178, 1980. DOI: 10.1016/0004-3702(80)90042-9 12

Ofra Amir and Ya'akov Gal. Plan recognition and visualization in exploratory learning environments. *ACM Transactions on Interactive Intelligent Systems (TIIS)*, 3(3):1–23, 2013. 86

Ofra Amir, Barbara J. Grosz, Edith Lok Man Law, and Roni Stern. Collaborative health care plan support. In *Proc. of the 12th International Conference on Autonomous Agents and Multiagent Systems*, ACM, 2013. DOI: 10.1145/2533670.2533674 86

John Anderson. *Langauge, Memory, and Thought*. Lawrence Erlbaum, 1976. DOI: 10.4324/9780203780954 8

Marcelo Gabriel Armentano and Analía Amandi. Plan recognition for interface agents. *Artificial Intelligence Review*, 28(2):131–162, 2007. DOI: 10.1007/s10462-009-9095-8 86

Dorit Avrahami-Zilberbrand and Gal A. Kaminka. Fast and complete symbolic plan recognition. In *IJCAI*, pages 653–658, 2005. 15, 17, 61, 62, 83, 84

Yoshua Bengio. From system 1 deep learning to system 2 deep learning. https://slideslive.com/38921750/from-system-1-deep-learning-to-system-2-deep-learning 85

F. Bisson, F. Kabanza, A. R. Benaskeur, and H. Irandoust. Provoking opponents to facilitate the recognition of their intentions. In *Proc. of the Conference of the Association for the Advancement of Artificial Intelligence (AAAI)*, 2011. 29, 86

Francis Bisson, Hugo Larochelle, and Froduald Kabanza. Using a recursive neural network to learn an agent's decision model for plan recognition. In *24th International Joint Conference on Artificial Intelligence*, 2015. 85

Nate Blaylock and James Allen. Recognizing instantiated goals using statistical methods. In *IJCAI Workshop on Modeling others from Observations (MOO)*, pages 79–86, 2005. 71, 83, 84

Thomas Bolander and Mikkel Birkegaard Andersen. Epistemic planning for single-and multi-agent systems. *Journal of Applied Non-Classical Logics*, 21(1):9–34, 2011. DOI: 10.3166/jancl.21.9-34 76

Michael E. Bratman. *Intention, Plans, and Practical Reason*. Harvard University Press, Cambridge, 1987. 7

Hung H. Bui, Svetha Venkatesh, and Geoff West. Policy recognition in the abstract hidden Markov model. *Journal of Artificial Intelligence Research*, 17:451–499, 2002. DOI: 10.1613/jair.839 24

Hung Hai Bui. A general model for online probabilistic plan recognition. In *IJCAI*, 3:1309–1315, Citeseer, 2003. 78, 79, 83, 84

Sandra Carberry. *Plan Recognition in Natural Language Dialogue*. ACL-MIT Press Series in Natural Language Processing, MIT Press, Cambridge, MA, 1990. 11

Tathagata Chakraborti, Sarath Sreedharan, Yu Zhang, and Subbarao Kambhampati. Plan explanations as model reconciliation: Moving beyond explanation as soliloquy. In *Proc. of the 26th International Joint Conference on Artificial Intelligence*, pages 156–163, 2017. DOI: 10.24963/ijcai.2017/23 86

Tathagata Chakraborti, Anagha Kulkarni, Sarath Sreedharan, David E. Smith, and Subbarao Kambhampati. Explicability? legibility? predictability? transparency? privacy? security? the emerging landscape of interpretable agent behavior. In *Proc. of the International Conference on Automated Planning and Scheduling*, 29:86–96, 2019. 86

Philip R. Cohen and Hector J. Levesque. Intention is choice with commitment, 42(2–3):213–261, 1990. DOI: 10.1016/0004-3702(90)90055-5 7

Cristina Conati, Abigail S. Gertner, Kurt VanLehn, and Marek J. Druzdzel. On-line student modeling for coached problem solving using Bayesian networks. In *Proc. of the 6th International Conference on User Modeling*, 1997. DOI: 10.1007/978-3-7091-2670-7_24 19, 86

Haskell Curry. *Foundations of Mathematical Logic*. Dover Publications Inc., 1977. 63

Johan de Kleer and Raymond Reiter. Foundations for assumption-based truth maintenance systems: Preliminary report. In *Proc. American Association for Artificial Intelligence National Conference*, pages 183–188, 1987. 81

Karel Durkota, Viliam Lisỳ, Branislav Bošanskỳ, and Christopher Kiekintveld. Optimal network security hardening using attack graph games. In *24th International Joint Conference on Artificial Intelligence*, 2015. 86

Yolanda E-Martín, María D. R.-Moreno, and David E. Smith. A fast goal recognition technique based on interaction estimates. In *Proc. of the 24th International Conference on Artificial Intelligence*, pages 761–768, 2015. 55, 57, 83, 84

Jay Earley. An efficient context-free parsing algorithm. *Communications of the ACM*, 13(2):94–102, 1970. DOI: 10.1145/362007.362035 60

Richard Fikes and Nils Nilsson. Strips: A new approach to the application of theorem proving to problem solving. *Artificial Intelligence*, 2:189–208, 1971. DOI: 10.1016/0004-3702(71)90010-5 3, 10

Jaime F. Fisac, Monica A. Gates, Jessica B. Hamrick, Chang Liu, Dylan Hadfield-Menell, Malayandi Palaniappan, Dhruv Malik, S. Shankar Sastry, Thomas L. Griffiths, and Anca D. Dragan. Pragmatic-pedagogic value alignment. In *Robotics Research*, pages 49–57, Springer, 2020. DOI: 10.1007/978-3-030-28619-4_7 75

Richard G. Freedman and Shlomo Zilberstein. Integration of planning with recognition for responsive interaction using classical planners. In *31st AAAI Conference on Artificial Intelligence*, 2017. 86

Richard G. Freedman, Hee-Tae Jung, and Shlomo Zilberstein. Plan and activity recognition from a topic modeling perspective. In *24th International Conference on Automated Planning and Scheduling*, 2014. 86

Richard G. Freedman, Yi Ren Fung, Roman Ganchin, and Shlomo Zilberstein. Responsive planning and recognition for closed-loop interaction. *AAAI Fall Symposium Series*, 2019. 86

Ya'akov Gal, Swapna Reddy, Andee Shieber, Stuart M. Rubin, and Barbara J. Grosz. Plan recognition in exploratory domains. *Artificial Intelligence*, 176(1):2270–2290, 2012. DOI: 10.1016/j.artint.2011.09.002 78, 83, 84, 86

Christopher Geib. Delaying commitment in probabilistic plan recognition using combinatory categorial grammars. In *Proc. IJCAI*, pages 1702–1707, 2009. 17, 61, 63, 83, 84

Christopher W. Geib and Robert P. Goldman. Plan recognition in intrusion detection systems. In *Proc. DARPA Information Survivability Conference and Exposition II (DISCEX'01)*, 1:46–55, IEEE, 2001. DOI: 10.1109/discex.2001.932191 78, 86

Christopher W. Geib and Robert P. Goldman. A probabilistic plan recognition algorithm based on plan tree grammars. *Artificial Intelligence*, 173(11):1101–1132, 2009a. http://dx.doi.org/10.1016/j.artint.2009.01.003 DOI: 10.1016/j.artint.2009.01.003 41

Christopher W. Geib and Robert P. Goldman. A probabilistic plan recognition algorithm based on plan tree grammars. *Artificial Intelligence*, 173(11):1101–1132, 2009b. DOI: 10.1016/j.artint.2009.01.003 32, 40

Christopher W. Geib and Pavan Kantharaju. Learning combinatory categorial grammars for plan recognition. In *32nd AAAI Conference on Artificial Intelligence*, 2018. 50, 85

C. W. Geib, J. Maraist, and R. P. Goldman. A new probabilistic plan recognition algorithm based on string rewriting. In *International Conference on Automated Planning and Scheduling (ICAPS)*, pages 91–98, 2008. 56, 58, 83, 84

Piotr J. Gmytrasiewicz and Prashant Doshi. A framework for sequential planning in multi-agent settings. *Journal of Artificial Intelligence Research*, 24:49–79, 2005. DOI: 10.1613/jair.1579 75, 83, 84

Robert P. Goldman and Eugene Charniak. Probabilistic text understanding. *Statistics and Computing*, 2(2):105–114, 1992. DOI: 10.1007/bf01889589 13

Ayelet Goldstein and Yuval Shahar. Implementation of a system for intelligent summarization of longitudinal clinical records. In *Process Support and Knowledge Representation in Health Care*, pages 68–82, Springer, 2013. DOI: 10.1007/978-3-319-03916-9_6 86

Roger Granada, Juarez Monteiro, Nathan Gavenski, and Felipe Meneguzzi. Object-based goal recognition using real-world data. In *Mexican International Conference on Artificial Intelligence*, pages 325–337, Springer, 2020. DOI: 10.1007/978-3-030-60884-2_24 86

Sheila Greibach. A new normal form theorem for context-free phrase structure grammars. *Journal of the ACM*, 12(1):42–52, 1965. DOI: 10.1145/321250.321254 40

Barbara J. Grosz. Discourse knowledge. In D. Walker, Ed., *Understanding Spoken language*, pages 229–347, North-Holland, NY, 1978. 12

Dylan Hadfield-Menell, Stuart J. Russell, Pieter Abbeel, and Anca Dragan. Cooperative inverse reinforcement learning. In *Advances in Neural Information Processing Systems*, pages 3909–3917, 2016. 86

Fritz Heider and Marianne Simmel. An experimental study of apparent behavior. *The American Journal of Psychology*, 57(2):243–259, 1944. DOI: 10.2307/1416950 1

Eric Horvitz, Jack Breese, David Heckerman, David Hovel, and Koos Rommelse. The lumiere project: Bayesian user modeling for inferring the goals and needs of software users. In *Proc. of the 14th Conference on Uncertainty in Artificial Intelligence*, 1998. 18, 86

Ronald Howard. *Dynamic Programming and Markov Processes*. MIT Press, 1960. 3

Yu-Sian Jiang, Garrett Warnell, and Peter Stone. Inferring user intention using gaze in vehicles. In *Proc. of the 20th ACM International Conference on Multimodal Interaction*, pages 298–306, 2018. DOI: 10.1145/3242969.3243018 86

A. Joshi and Y. Schabes. Tree-adjoining grammars. In *Handbook of Formal Languages*, 3:69–124, Springer Verlag, 1997. DOI: 10.1007/978-3-642-59126-6_2 40

F. Kabanza, J. Filion, A. R. Benaskeur, and H. Irandoust. Controlling the hypothesis space in probabilistic plan recognition. In *IJCAI*, pages 2306–2312, 2013. 66, 83, 84

Leslie Kaelbling. Doing for our robots what nature did for us. 2019. https://smartech.gatech.edu/handle/1853/60966 85

Pavan Kantharaju, Santiago Ontanón, and Christopher W. Geib. Scaling up CCG-based plan recognition via Monte-Carlo tree search. In *IEEE Conference on Games (CoG)*, pages 1–8, IEEE, 2019. DOI: 10.1109/cig.2019.8848013 66, 67, 83, 84, 85

Henry Kautz and James F. Allen. Generalized plan recognition. In *Proc. of the Conference of the American Association of Artificial Intelligence (AAAI)*, pages 32–38, 1986. 9, 10, 27

Sarah Keren, Avigdor Gal, and Erez Karpas. Goal recognition design. In *24th International Conference on Automated Planning and Scheduling*, 2014. 29, 70, 71

Sarah Keren, Avigdor Gal, and Erez Karpas. Goal recognition design in deterministic environments. *Journal of Artificial Intelligence Research*, 65:209–269, 2019. DOI: 10.1613/jair.1.11551 29, 70

Sarah Keren, Avigdor Gal, and Erez Karpas. Goal recognition design—survey, *Proc. of the 29th International Joint Conference on Artificial Intelligence (IJCAI-20)*, Survey Track, 2020. DOI: 10.24963/ijcai.2020/675 29, 70

Joseph Kim, Matthew E. Woicik, Matthew C. Gombolay, Sung-Hyun Son, and Julie A. Shah. Learning to infer final plans in human team planning. In *IJCAI*, pages 4771–4779, 2018. DOI: 10.24963/ijcai.2018/663 85

Birgit Knudsen and Ulf Liszkowski. Eighteen- and 24-month-old infants correct others in anticipation of action mistakes. *Developmental Science*, 15:113–22, January 2012. DOI: 10.1111/j.1467-7687.2011.01098.x 1

Donald E. Knuth, James H. Morris, and Vaughan R. Pratt. Fast pattern matching in strings*. *SIAM Journal on Computing*, 6(2):323–350, 1977. DOI: 10.1137/0206024 13

John Laird. *The Soar Cognitive Architecture*. MIT Press, 2012. DOI: 10.7551/mitpress/7688.001.0001 8

Kennard Laviers, Gita Sukthankar, David Aha, and Matthew Molineaux. Improving offensive performance through opponent modeling. In *Proc. of the AAAI Conference on Artificial Intelligence and Interactive Digital Entertainment*, 4(1), 2009. 21, 86

Steven James Levine and Brian Charles Williams. Concurrent plan recognition and execution for human-robot teams. In *24th International Conference on Automated Planning and Scheduling*, 2014. 8, 25, 78, 80, 83, 84, 86

Lin Liao, Dieter Fox, and Henry A. Kautz. Location-based activity recognition using relational Markov networks. In *Proc. of IJCAI*, pages 773–778, 2005. 22, 23, 24, 86

Rudolf Lioutikov, Guilherme Maeda, Filipe Veiga, Kristian Kersting, and Jan Peters. Learning attribute grammars for movement primitive sequencing. *The International Journal of Robotics Research*, 39(1):21–38, 2020. DOI: 10.1177/0278364919868279 85

Viliam Lisỳ, Radek Píbil, Jan Stiborek, Branislav Bošanskỳ, and Michal Pĕchouček. Game-theoretic approach to adversarial plan recognition. In *Proc. of the 20th European Conference on Artificial Intelligence*, pages 546–551, IOS Press, 2012. 78, 79, 83, 84, 86

Diane J. Litman and James F. Allen. A plan recognition model for subdialogues in conversations. *Cognitive Science*, 11(2):163–200, 1987. DOI: 10.1207/s15516709cog1102_4 12

J. Maraist. String shuffling over a gap between parsing and plan recognition. *AAAI Workshop on Plan Activity and Intent Recognition*, 2017. 56, 59, 83, 84

P. Masters and S. Sardina. Cost-based goal recognition for path-planning. In *AAMAS*, pages 750–758, International Foundation for Autonomous Agents and Multiagent Systems, 2017. 55, 56, 83, 84

Mariane Maynard, Thibault Duhamel, and Froduald Kabanza. Cost-based goal recognition meets deep learning. *ArXiv Preprint ArXiv:1911.10074*, 2019. 85

John McCarthy and Patrick J. Hayes. Some philosophical problems from the standpoint of artificial intelligence. In B. Meltzer and D. Michie, Eds., *Machine Intelligence*, vol. 4, Edinburgh University Press, Edinburgh, 1969. DOI: 10.1016/b978-0-934613-03-3.50033-7 12

Drew McDermott, Malik Ghallab, Adele Howe, Craig Knoblock, Ashwin Ram, Manuela Veloso, Daniel Weld, and David Wilkins. PDDL—the planning domain definition language. *Technical Report CVC TR98003/DCS TR1165*, New Haven, CT, Yale Center for Computational Vision and Control, 1998. 44

R. Mirsky and Y. Gal. Slim: Semi-lazy inference mechanism for plan recognition. In *International Joint Conference of Artificial Intelligence (IJCAI)*, 2016. 61, 65, 83, 84

R. Mirsky, Y. Gal, and S. M. Shieber. Cradle: An online plan recognition algorithm for exploratory domains. *ACM Transactions on Intelligent Systems and Technology (TIST)*, 8(3):45–1, 2017a. DOI: 10.1145/2996200 66, 77, 83, 84

Reuth Mirsky, Ya'akov Gal, and David Tolpin. Session analysis using plan recognition. *Workshops at the 31st AAAI Conference on Artificial Intelligence*, 2017b. 86

Reuth Mirsky, Roni Stern, Kobi Gal, and Meir Kalech. Sequential plan recognition: An iterative approach to disambiguating between hypotheses. *Artificial Intelligence*, 260:51–73, 2018. 29, 74, 83, 84, 86

Reuth Mirsky, Kobi Gal, Roni Stern, and Meir Kalech. Goal and plan recognition design for plan libraries. *ACM Transactions on Intelligent Systems and Technology (TIST)*, 10(2):1–23, 2019a. 29, 70, 71

Reuth Mirsky, Ahmad Majadly, Kobi Gal, Rami Puzis, Ariel Felner, et al. New goal recognition algorithms using attack graphs. In *International Symposium on Cyber Security Cryptography and Machine Learning*, pages 260–278, Springer, 2019b. DOI: 10.1007/978-3-030-20951-3_23 86

Reuth Mirsky, William E. Macke, Andy Wang, Harel Yedidsion, and Peter Stone. A penny for your thoughts: The value of communication in ad hoc teamwork. In *IJCAI*, 2020. DOI: 10.24963/ijcai.2020/36

Christian J. Muise, Vaishak Belle, Paolo Felli, Sheila A. McIlraith, Tim Miller, Adrian R. Pearce, and Liz Sonenberg. Planning over multi-agent epistemic states: A classical planning approach. In *Proc. of the AAAI Conference on Artificial Intelligence*, 29(1):3327–3334, 2015. 76

Yoshifumi Murakami, Yoshinori Kuno, Nobutaka Shimada, and Yoshiaki Shirai. Collision avoidance by observing pedestrians' faces for intelligent wheelchairs. *Journal of the Robotics Society of Japan*, 20(2):206–213, 2002. 86

Allen Newell, J. C. Shaw, and Herbert A. Simon. Report on the general problem solver. In *Proc. of the International Conference on Information Processing*, pages 256–264, 1959. 3, 4

Minjing Peng, Yanwei Qin, Chenxin Tang, and Xiangming Deng. An e-commerce customer service robot based on intention recognition model. *Journal of Electronic Commerce in Organizations (JECO)*, 14(1):34–44, 2016. DOI: 10.4018/jeco.2016010104 86

Ramon Fraga Pereira, Nir Oren, and Felipe Meneguzzi. Landmark-based heuristics for goal recognition. In *31st AAAI Conference on Artificial Intelligence (AAAI-17)*, AAAI Press, 2017. DOI: 10.1016/j.artint.2019.103217 66, 68

Ramon Fraga Pereira, Mor Vered, Felipe Meneguzzi, and Miquel Ramírez. Online probabilistic goal recognition over nominal models. *IJCAI*, pages 5547–5553, 2019. 83, 84, 85

Ronald P. A. Petrick and Fahiem Bacchus. A Knowledge-Based Approach to Planning with Incomplete Information and Sensing. In *AIPS*, 2:212–222, 2002. 76

Martha E. Pollack. A model of plan inference that distinguishes between the beliefs of actors and observers. In *Proc. of the 24th Annual Meeting on Association for Computational Linguistics*, pages 207–214, Association for Computational Linguistics, 1986. DOI: 10.3115/981131.981160 75

Martha E. Pollack. The uses of plans. *Artificial Intelligence*, 57(1):43–68, September 1992. DOI: 10.1016/0004-3702(92)90104-6 12

David V. Pynadath and Michael P. Wellman. Accounting for context in plan recognition, with application to traffic monitoring. In *Proc. of the 11th Conference on Uncertainty in Artificial Intelligence*, pages 472–481, 1995. 14, 16

Xinzhou Qin and Wenke Lee. Attack plan recognition and prediction using causal networks. In *20th Annual Computer Security Applications Conference*, pages 370–379, IEEE, 2004. DOI: 10.1109/csac.2004.7 86

M. Ramırez and H. Geffner. Probabilistic plan recognition using off-the-shelf classical planners. In *Proc. of the Conference of the Association for the Advancement of Artificial Intelligence (AAAI)*, Citeseer, 2010. 43, 49, 83, 84

Miquel Ramírez and Hector Geffner. Plan recognition as planning. In *21st International Joint Conference on Artificial Intelligence*, 2009. 20, 27, 32, 47, 48, 57, 84

Patrice Roy, Bruno Bouchard, Abdenour Bouzouane, and Sylvain Giroux. A hybrid plan recognition model for alzheimer's patients: Interleaved-erroneous dilemma. *Web Intelligence and Agent Systems: An International Journal*, 7(4):375–397, 2009. DOI: 10.3233/wia-2009-0175 86

Stuart Russell and Peter Norvig. *Artificial Intelligence: A Modern Approach*, 2nd ed., Prentice-Hall, 2003. DOI: 10.1093/oso/9780190905033.003.0012 7, 23, 78

Earl Sacerdoti. The nonlinear nature of plans. In *Proc. 4th IJCAI*, pages 206–214, 1975. 10

Akanksha Saran, Srinjoy Majumdar, Elaine Schaertl Short, Andrea Thomaz, and Scott Niekum. Human gaze following for human-robot interaction. In *IEEE/RSJ International Conference on Intelligent Robots and Systems (IROS)*, pages 8615–8621, 2018. DOI: 10.1109/iros.2018.8593580 86

C. F. Schmidt, N. S. Sridharan, and J. L. Goodson. The plan recognition problem: An intersection of psychology and artificial intelligence. *Artificial Intelligence*, 11:45–83, 1978. DOI: 10.1016/0004-3702(78)90012-7 3, 4

José Á Segura-Muros, Raúl Pérez, and Juan Fernández-Olivares. Using inductive rule learning techniques to learn planning domains. In *International Conference on Information Processing and Management of Uncertainty in Knowledge-Based Systems*, pages 642–656, Springer, 2018. DOI: 10.1007/978-3-319-91479-4_53 85

Michael Shum, Max Kleiman-Weiner, Michael L. Littman, and Joshua B. Tenenbaum. Theory of minds: Understanding behavior in groups through inverse planning. In *Proc. of the AAAI Conference on Artificial Intelligence*, 33:6163–6170, 2019. DOI: 10.1609/aaai.v33i01.33016163 86

Maayan Shvo and Sheila A. McIlraith. Active goal recognition. In *Proc. of the Conference of the Association for the Advancement of Artificial Intelligence (AAAI)*, 2020a. DOI: 10.1609/aaai.v34i06.6551 29, 86

Maayan Shvo and Sheila A. McIlraith. Active goal recognition. In *AAAI*, pages 9957–9966, 2020b. DOI: 10.1609/aaai.v34i06.6551 74, 83, 84

Maayan Shvo, Toryn Klassen, Shirin Sohrabi, and Sheila McIlraith. Epistemic plan recognition. In *Proc. of the 19th Conference on Autonomous Agents and Multi-Agent Systems (AAMAS)*, 2020. 75, 83, 84

Ronal Singh, Tim Miller, Joshua Newn, Liz Sonenberg, Eduardo Velloso, and Frank Vetere. Combining planning with gaze for online human intention recognition. In *Proc. of the 17th International Conference on Autonomous Agents and Multiagent Systems*, pages 488–496, 2018. 74, 83, 84

S. Sohrabi, A. V. Riabov, and O. Udrea. Plan recognition as planning revisited. In *IJCAI*, pages 3258–3264, 2016. 72, 83, 84

Sarath Sreedharan, Subbarao Kambhampati, et al. Handling model uncertainty and multiplicity in explanations via model reconciliation. In *28th International Conference on Automated Planning and Scheduling*, 2018. 86

Mark Steedman. *The Syntactic Process*. MIT Press, 2000. DOI: 10.7551/mitpress/6591.001.0001 63

Nathan R. Sturtevant. Benchmarks for grid-based pathfinding. *IEEE Transactions on Computational Intelligence and AI in Games*, 4(2):144–148, 2012. DOI: 10.1109/tciaig.2012.2197681 56

Gita Sukthankar and Katia Sycara. A cost minimization approach to human behavior recognition. In *Proc. of the 4th International Joint Conference on Autonomous Agents and Multiagent Systems*, pages 1067–1074, 2005. 86

Gita Sukthankar and Katia P. Sycara. Hypothesis pruning and ranking for large plan recognition problems. In *AAAI*, 8:998–1003, 2008. 86

Shao-Hua Sun, Te-Lin Wu, and Joseph J. Lim. Program guided agent. In *International Conference on Learning Representations*, 2020. https://openreview.net/forum?id=BkxUvnEYDH

Gabriel Synnaeve and Pierre Bessiere. A Bayesian model for plan recognition in RTS games applied to starcraft. In *7th Artificial Intelligence and Interactive Digital Entertainment Conference*, 2011. 86

Kartik Talamadupula, Gordon Briggs, Tathagata Chakraborti, Matthias Scheutz, and Subbarao Kambhampati. Coordination in human-robot teams using mental modeling and plan recognition. In *IEEE/RSJ International Conference on Intelligent Robots and Systems*, pages 2957–2962, 2014. DOI: 10.1109/iros.2014.6942970 86

Austin Tate. Generating project networks. In *IJCAI*, pages 888–893, 1977. 10

M. Vered and G. A. Kaminka. Heuristic online goal recognition in continuous domains. In *IJCAI*, pages 4447–4454, 2017. 73, 74, 83, 84

Marc B. Vilain. Getting serious about parsing plans: A grammatical analysis of plan recognition. In *Proc. AAAI*, pages 190–197, 1990. 13

Felix Warneken and Michael Tomasello. Helping and cooperation at 14 months of age. *Infancy*, 11:271–294, May 2007. DOI: 10.1111/j.1532-7078.2007.tb00227.x 1

Felix Warneken and Michael Tomasello. The roots of human altruism. *British Journal of Psychology (London, England: 1953)*, 100:455–71, August 2009. DOI: 10.1348/000712608x379061 1

C. Wayllace, P. Hou, W. Yeoh, and T. C. Son. Goal recognition design with stochastic agent action outcomes. In *IJCAI*, 2016. 70

C. Wayllace, P. Hou, and W. Yeoh. New metrics and algorithms for stochastic goal recognition design problems. In *IJCAI*, 2017. DOI: 10.24963/ijcai.2017/622 29

Christabel Wayllace, Sarah Keren, William Yeoh, Avigdor Gal, and Erez Karpas. Accounting for partial observability in stochastic goal recognition design: Messing with the marauder's map. In *Proc. of the European Conference on Artificial Intelligence*, 2020. 29, 70

Daniel Younger. Recognition and parsing of context-free languages in time n^3. *Information and Control*, 2(10):189–208, 1967. DOI: 10.1016/s0019-9958(67)80007-x 16

Authors' Biographies

REUTH MIRSKY

Reuth Mirsky is a postdoctoral fellow at the Computer Science department in the University of Texas at Austin, under the mentorship of Prof. Peter Stone. Reuth received her Ph.D. from the Department of Software and Information Systems Engineering at Ben Gurion University, under the supervision of Prof. Kobi Gal. Her Ph.D. thesis was on plan recognition in exploratory environments. Reuth's research focuses on improving existing AI with human-inspired design, and her algorithms have been applied in various tasks for education, clinical treatment, and finance. Reuth's contributions were published by leading AI and HRI conferences and journals. Her long-term research vision is to enable better collaborations in mixed human-and-artificial agents settings.

SARAH KEREN

Sarah Keren is a postdoctoral fellow at Harvard University, where she is affiliated with the Center for Research on Computation and Society (CRCS). Her mentors are Prof. Barbara Grosz and Prof. David Parkes. Before coming to Harvard, Sarah completed her Ph.D. at the Faculty of Industrial Engineering and Management of the Technion–Israel Institute of Technology, where she was advised by Prof. Avigdor Gal and Dr. Erez Karpas. Sarah's research focuses on manipulating and redesigning environments for optimizing their utility. In particular, her Ph.D. work established the task of Goal Recognition Design, where environments are manipulated to maximize the ability to recognize the goals of agents acting within them. Sarah's work has appeared in three leading artificial intelligence conferences (AAAI, ICAPS, and IJCAI). She has received various excellence awards, including an honorable mention for best paper at ICAPS 2014 as well as the Eric and Wendy Schmidt Postdoctoral Award for Women in Mathematical and Computing Sciences.

CHRISTOPHER GEIB

Christopher Geib is a Principal Researcher at SIFT LLC and an internationally recognized researcher in probabilistic plan recognition and planning. He received his Ph.D. in Computer Science from the University of Pennsylvania in 1995. Prior to joining SIFT, he had an extensive career both in academia as an Associate Professor at Drexel University and a Research Fellow at the University of Edinburgh, and in industry as a Principal Research Scientist at Honeywell. He has published more than 50 scholarly publications. His interests include probabilistic plan recognition and planning under uncertainty based on formal grammars and interaction between human and synthetic agents using actions and language. He has been the principal architect of multiple plan recognition systems over the last 20+ years.

Printed in the United States
by Baker & Taylor Publisher Services